# THE GROWTH OF MEDICAL KNOWLEDGE

PHILOSOPHY AND MEDICINE



VOLUME 36

# THE GROWTH OF MEDICAL KNOWLEDGE

*Edited by*

HENK A. M. J. TEN HAVE

*University of Limburg, Faculty of Health Sciences, Maastricht, Netherlands*

GERRIT K. KIMSMA

*Free University, Amsterdam, Netherlands*

STUART F. SPICKER

*School of Medicine, University of Connecticut, Health Center, Farmington, Connecticut, U.S.A.*

KLUWER ACADEMIC PUBLISHERS

DORDRECHT / BOSTON / LONDON

**Library of Congress Cataloging-in-Publication Data**

The Growth of medical knowledge / edited by Henk ten Have, Gerrit
Kimsma, Stuart F. Spicker.
p. cm. -- (Philosophy and medicine ; v. 36)
Based on a conference held in Maastricht, the Netherlands in 1987;
sponsored by the European Society for Philosophy of Medicine and
Health Care (ESPMH).
ISBN 0-7923-0736-4 (U.S. : alk. paper)
1. Medicine--Philosophy--Congresses. 2. Medicine--Research-
-Congresses. 3. Medical care--Congresses. I. Have, H. ten.
II. Kimsma, Gerrit. III. Spicker, Stuart F., 1937- .
IV. European Society for Philosophy of Medicine and Health Care.
V. Series.
[DNLM: 1. Anthroposophy--congresses. 2. Philosophy, Medical-
-congresses. W3 PH609 v. 36 / W 61 G884 1987]
R723.G76 1990
610'.1--dc20
DNLM/DLC
for Library of Congress 90-4350

ISBN 0-7923-0736-4

---

Published by Kluwer Academic Publishers,
P.O. Box 17, 3300 AA Dordrecht, The Netherlands.

Kluwer Academic Publishers incorporates
the publishing programmes of
D. Reidel, Martinus Nijhoff, Dr W. Junk and MTP Press.

Sold and distributed in the U.S.A. and Canada
by Kluwer Academic Publishers Group,
101 Philip Drive, Norwell, MA 02061, U.S.A.

In all other countries, sold and distributed
by Kluwer Academic Publishers,
P.O. Box 322, 3300 AH Dordrecht, The Netherlands.

*printed on acid-free paper*



Printed in The Netherlands

## TABLE OF CONTENTS

## SECTION III / IMAGE OF MAN AND THE GROWTH OF MEDICAL KNOWLEDGE

# PREFACE

The growth of knowledge and its effects on the practice of medicine have been issues of philosophical and ethical interest for several decades and will remain so for many years to come. The outline of the present volume was conceived nearly three years ago. In 1987, a conference on this theme was held in Maastricht, the Netherlands, on the occasion of the founding of the European Society for Philosophy of Medicine and Health Care (ESPMH). Most of the chapters of this book are derived from papers presented at that meeting, and for the purpose of editing the book Stuart Spicker, Ph.D., joined two founding members of ESPMH, Henk ten Have and Gerrit Kimsma. The three of them successfully brought together a number of interesting contributions to the theme, and ESPMH is grateful and proud to have initiated the production of this volume.

The Society intends that annual meetings be held in different European countries on a rotating basis and to publish volumes related to these meetings whenever feasible. In 1988, the second conference was held in Aarhus, Denmark on "Values in Medical Decision Making and Resource Allocation in Health Care". In 1989, a meeting was held in Czestochowa, Poland, on "European Traditions in Philosophy of Medicine. From Brentano to Bieganski". It is hoped that these conferences and the books to be derived from them, will initiate a new European tradition, lasting well into the 21st century!

P. J. THUNG,
*ESPMH President, 1987–1989*

HENK A.M.J. TEN HAVE AND STUART F. SPICKER

# INTRODUCTION

In his introductory courses, W. Arens did what most professors of anthropology used to do. He was lecturing on kinship, politics and economics until one of his students asked why he did not pay attention to a more interesting topic, e.g., cannibalism. This question prompted Arens to study man-eating, resulting in the publication of *The Man-Eating Myth* in 1979 [1]. After assessing critically the instances of and documentation for cannibalism, Arens concluded that there is no satisfactory evidence of the existence of anthropophagy as a socially approved custom in any part of the world: the idea of the cannibalistic nature of *homo sapiens sapiens* is a myth. When the evidence from all fields on the world's man-eaters is so sparse, how can we explain the innumerable literary references alluding to cannibalism, and particularly the preoccupation of professional anthropologists with describing and interpreting it as relatively commonplace?

The real problem, perhaps, is not so much the mythical nature of anthropophagy but the use and function of this fiction within a modern science like anthropology.

By rephrasing the problem, Arens raised an important issue for science in general. Science cannot evolve without prior beliefs or basic assumptions that become easily transformed into facts and incorporated within the conventional wisdom. Without at least some presuppositions scientific activity would cease. The myth of man-eating evolved from presuppositions concerning human nature which are constitutive of modern anthropology. Anthropology creates a conceptual order based on differences between civilized and savage modes of existence. Without assuming that others are categorically different from ourselves, anthropology could not find its own identity. Not only is the acceptance of such differences, and the avoidance of moral judgments on the value of variations among cultures (i.e., cultural relativism) basic to the emergence of anthropology as a science, but to understand the nature of humanity throughout Western civilization anthropologists needed exotic images of barbarism and the collapse of cultures. In other words, cannibalism is used to interpret a culture's own identity vis-à-vis its neighboring societies or to mark the progress of a particular culture. Gradually this popular notion of savagery became transformed into a description of the



*H.A.M.J. ten Have et al. (eds.), The Growth of Medical Knowledge, 1–11.*

human condition. Within the everyday social world such a notion is primarily evaluative: it is used to construct a dichotomy between "we" and "they." Within the world of science, cannibalism is taken as an empirical datum in need of an explanation. Various theories were proposed to explain why various cultures condoned and practiced eating human flesh.

Arens' analysis is important because it brings anthropology one step further by formulating a meta-anthropological problem: Why does one culture assume that other cultures are anthropophagic? Taking a critical and reflexive attitude towards the discipline itself, Arens probed to a further understanding. Whether anthropophagy is fact or fiction, of interest is the almost universal prevalence of this belief; i.e., it has been useful in different cultures and in anthropology itself. The proper topic of anthropological research is the variety of social and cultural ways the world has been constructed and interpreted.

Modern philosophy of science, too, has learned a lesson from quantum physics: The idea of an objective reality or natural order was replaced by *the representation* of our knowledge of this reality. More and more mankind is confronted not with reality but with its own construction of reality.

From this digression to anthropology three conclusions relevant to medicine may be drawn:

1. Scientific activity begins with presuppositions concerning nature, reality, and humanity. These presuppositions themselves are not usually questioned, but are commonly accepted as the foundation upon which to erect a discipline.
2. Because certain matters are in fact generally presupposed, scientific activity becomes possible, constant controversy over fundamentals can be avoided, and research activities can be linked and coordinated.
3. Science can advance in at least two different ways: (1) It can evolve by attending to the basis of conventional and accepted presuppositions, or (2) it can progress by questioning and challenging what it has simply presupposed. Science progresses, in Kuhnian terms, by working within an established scientific tradition governed by *paradigms,* or by more radical revolutions that result in the adoption of new paradigms [11].

With these thoughts in mind, the theme of this volume, "The Growth of Medical Knowledge," immediately raises a number of questions. Is there such a thing as *progress* in medicine, and, if so, how should we understand it? In what way is knowledge important to medicine? What kind of human endeavor is medicine? If it is a science and if it progresses, is the philosophy of science of any use to it? These questions refer to central problems in the philosophy of medicine. They all seem to converge at one fundamental question: "What is medicine?" To understand the activity in which they are involved, to explicate the basic meaning of the profession they have chosen, or to clarify what characterizes their theories and actions as 'medical', physicians have always philosophized on their *officium nobile*. Most of the time medicine has situated itself "between two cultures" – articulated in various ways – science versus art, theory versus praxis, speculation versus empiricism, formalism versus intuitionism, reductionism versus holism, analysis versus synthesis, *Naturwissenschaften* versus *Geisteswissenschaften*.

Two strategies that have been employed to eliminate these dichotomies we shall label the "reduction strategy" and the "autonomy strategy."

### *A. Reduction Strategy*

When adopting the reduction strategy one set of characteristics is reduced to an opposing set or considered a deviation; it may even be shown to be irrelevant, so that, for example, medicine can be fully described in terms of these particular characteristics. Thus an apparent dichotomy can be judged a mere misunderstanding upon further reflection. The dual nature of medicine, its Janus face, is deceptive. One aim of the philosophy of medicine is to clarify what is properly meant by the term 'medicine'.

This reductive strategy has resulted in two restricted and mutually exclusive conceptions of medicine:

(1) Ever since the famous dictum of the nineteenth-century German clinician, Bernhard Naunyn – "medicine shall be a natural science or shall not be" – medicine has been defined either as an autonomous science or was subsumed under another science. Medicine, of course, investigates the empirical phenomena of disease, attempts to establish laws that govern these phenomena, and tests theories and explanations through diagnoses and therapies. This view has recently been defended by L.A. Forstrom. He argues that clinical medicine conforms to what is ordinarily understood by "science." Clinical medicine is a science but it is also distinct from other sciences since (1) it has a specific natural domain ("... the human organism, in its manifold environmental contexts, in health and disease"), and (2) it has a specific investigative function ("... the formulation of concepts and generalizations ... arising from the clinician's systematic approach to phenomena as they occur in intact, living human organisms") ([6], pp. 9,11).

(2) A completely different argument is that medicine is primarily or

merely an art (*ars medendi*). This view was quite influential during the first decades of this century, and was defended by German medical philosophers such as Ernst Schweninger, Richard Koch, and Georg Honigmann [20]. Medicine is, for them, synonymous with *Krankenbehandlung*, the treatment of sick persons. Science is of no relevance to the *practice* of health care, it is the artistic capability to heal patients (*Kunstfertigkeit*) that is constitutive of medicine, not, for example, the sciences of physiology, pathology, and pharmacology.

Although both conceptions of medicine continue to remain within the professionals' self-image, it has become less plausible to regard medicine as either exclusively a science or an art. Medicine is so pervasive, so powerful, and so firmly established in modern societies, that there is no longer any professional effort to interpret and promote medicine as a science or art.

More importantly, the reductive strategy presupposes a specific relationship between thought and action, idea and practice: theoretical knowledge is first; it is subsequently "applied." Laboratory research produces scientific knowledge that finds its application in the clinic. The growing body of knowledge concerning microbes and the rise of sciences like bacteriology, virology and medical genetics have significantly affected the practice of modern medicine.

The assumed priority of theoretical factors has led to a tradition of historiography known as "Whig history" [4]. History is described retrospectively as an evolutionary process, progressing towards the present situation of the historian. The study of the history of medicine, for instance, has for a long time been guided by the belief that medical practice must be based on an understanding of the structure and functions of the body and the pathological processes that affect it. Therefore the history of medicine focused on discoveries, "great men" and powerful movements. The progress in medical practice was accounted for by the growth of knowledge in the various basic sciences. This approach to the history of medicine has been criticized in recent years from within historiography itself, particularly due to the increasing interest in social history [8, 19]. Detailed case studies made clear that advances in scientific knowledge were in many cases not directly related to therapeutic successes, decreased mortality, or the improved health status of a given population. For instance, Paul Ehrlich's discovery of Salvarsan in 1909, the first effective treatment for syphilis, did not eliminate venereal diseases [3]. The availability of knowledge as such is not enough; to make it effective, it must be attuned to specific health care practices through the creation of a network of facilities, services, institutions, and regulations. The social and cultural context must be transformed in order to make bacteriological or chemotherapeutic knowledge effective. Instead of studying the activities of the medical profession or the production of biomedical scientific knowledge, attention should be given to the socio-cultural milieu. The same point is made in the well-known work of McKeown: major improvements in health are not typically the result of biomedical research and better knowledge of infectious diseases, but they are associated with changes in the social environment such as better nutrition, improved hygiene, control of water supplies [15].

Thus, recent historical studies emphasize that additional scientific knowledge does not automatically lead to more efficacious and novel medical practices. The traditional belief in progress is rejected by the relativistic perspective. Sometimes there is nothing more than an accumulation of data and knowledge without any significant influence on practice. At other times there is a substantial growth of knowledge but its application is slow and even limited. Yet practices do change and have changed for the better. Using criteria such as mortality and morbidity indices, age, and health status to evaluate medical care, the progress of medicine seems undeniable. Still, it is clear from some recent studies that the relationship between knowledge and practice, as well as the role of medicine in its socio-cultural context, is more complicated and problematic than had been previously assumed in the history of medicine and in various *reductive* philosophies of medicine. If there is any progress in medicine, it cannot be accounted for solely by pointing to the growth of empirical knowledge. The development of more sophisticated insights do, however, raise an intriguing question: In what respect are these insights more advanced than those they reject? It seems as if there is a similar evolution to the one observed in anthropology. From philosophical and historical studies we know more about the ways medicine structures the world and socially constructs medical practices. Not medicine itself, but medical concepts are quite important. Such a shift of theoretical emphasis illustrates that at least some ideas can affect actual practices.

Such questions become more pressing when we consider another presupposition in the reductive strategy: The distinction between disease and illness. The growth of medical knowledge is usually connected with an increased understanding of diseases as biological or psychophysiological entities diagnosed by physicians. The entities are considered universal and independent of any socio-cultural context. They are the true reality, the underlying structure that specifies and can explain the experiences of patients. Physicians have acquired the ability to "read" the surface phenomena of the individual patient's illness and to relate them to this basic structure. To the degree that diseases reflect the *ontological* order, medical knowledge is judged to be in a more advanced state.

Focusing on the relationship of medical knowledge to changing socio-cultural contexts, the concept of disease is considered not as a representation of empirical reality but as a social construct [7]. What is experienced as illness, pain, or dysfunction, is given a meaning by constructing it into an object of therapeutic attention. Disease, therefore, is not a fact of nature but the result of socially and historically determined interpretive processes. If diseases are social constructs, we can better understand why there has been such great variation among disease concepts. Since different meanings are attributed to problems and complaints in different historical and social contexts, we have become aware of the diachronic and synchronic varieties of disease constructs.

Medicine is not the universal science or art it pretends to be. Medical practices differ widely, even in closely related Western nations [16]. In different countries different diseases are diagnosed; for instance, spasmophilia and liver crisis in France, or vasovegetative dystonia in West Germany. These findings make it extremely difficult to judge what counts for progress in medicine. What criterion can we use when medical practices seem so deeply affected by socio-cultural contexts? The answer is usually given by taking science as such as a sign of progress. For example, the substitution of a medical for a moral view on venereal disease has gradually taken place since 1880. But this view is problematic. It is not only questionable whether a medical interpretation is more effective than a moral one, but it suggests a different approach to the problem and to new practices that are not necessarily followed by better results [3]. Moreover, both interpretations are normative; they invoke and reinforce values, though different ones. How can it be argued that medicalization is better than moralization? If there is any progress it seems to be in the reconciliation of these perspectives. The variety of medical practices in different countries makes us aware of options that exist beyond our cultural borders. For example, Arens' analysis of anthropophagy enables us to acquire greater insight into our own constructive activities. This strategy brings us not only to the discovery of a wider range of options, initially assumed unavailable, but it leads also to a higher level of understanding of our own self-experience.

### *B. Autonomy Strategy*

Partly in response to the above criticisms, a second strategy has been employed to characterize medicine. Instead of seeking to reduce one set of characteristics to the other, medicine is identified as an *autonomous* discipline that functions between two cultures, possessing components of both but not being fully describable in terms of either. Although this strategy is popular nowadays, it had many prior advocates. For instance, Kneucker, in his 1949 monograph on philosophy of medicine [10], argued that medicine is a modified science situated between the natural sciences and the humanities. In medicine, knowledge is not an end in itself but only a means to treatment; its object is not an abstract disease but an individual patient. Appreciating this characteristic of medicine provides fundamental knowledge at a meta-level.

A few years earlier, Inlow, his American colleague, had written that medicine "... though made up of many parts, and making use of the findings and techniques of many sciences, is nevertheless an individual and organic whole in its own right" ([9], p. 273). It is an autonomous field of human endeavor with its own principles, logic, language, methods, and rules. The growth of scientific knowledge clearly had a positive influence on medicine, but the relationship is far more complicated than generally thought, even this growth was not always associated with improved medical practice. So when medicine is conceived as a separate discipline it is difficult to say whether it has progressed, without specifying the complex ways in which it is different from other applied sciences. Precisely this issue appears central in present-day philosophy of medicine [18].

We have witnessed debates concerning the biomedical model – whether, for example, medicine is a science of individuals, or possesses scientific status given its clinical knowledge and its applications. What these debates made clear is how problematic the current concept of science and the theory/practice distinction are in the realm of medicine. More and more, medicine is characterized as a practical science [21]. Its goal is not primarily to *know* but to *act* and intervene within the natural and social worlds. The aims of medicine and the theoretical sciences are different, but as a practical science medicine is not merely the application of the cognitive results of theoretical sciences. Praxis poses its own problems and questions. Although medicine employs these theories, its practical requirements determine the choice of treatments. Medicine is constituted through actions that are carefully reviewed, planned, and rationally motivated. Between theoretical knowledge and practical action there is no logical relationship. Nomological statements produced by scientific research are a heuristic device for the development of nomological statements ("If treatment X is given to patients with disease M, then M is cured") which guide the formulation of technological rules ("To cure a patient with disease M, treatment X should be given"). These rules have no truth-value but rather an effectiveness-value. The truth of nomological statements does not guarantee the effectiveness of the corresponding technological rules [17].

The intricate relation that obtains between theory and practice, generalization and unique reference, and truth and effectiveness within the practical science of medicine can itself be studied at a theoretical level. Instead of formulating "grand theories," philosophy of medicine should also analyze *in detail* daily health care practices. For this type of analysis, traditional distinctions between philosophy, sociology, anthropology, and history of medicine seem no longer relevant. In fact, such reorientation corresponds with the empirical turn in contemporary philosophy of science [2].

Since Kuhn, three issues have been emphasized in science research.

(1) *The social character of science.* Here science is primarily regarded as a collective affair, an activity of *scientific communities* with shared commitments and a common paradigm or disciplinary matrix. Under this construction, cognitive and social factors are closely related. These notions were described by L. Fleck in 1935, and only recently rediscovered.

(2) *The dynamic character of scientific knowledge.* The development of scientific knowledge is not cumulative, not a linear progression, but the result of a competition between different paradigms. New ideas, concepts, and theories replace older ones. "Growth" of knowledge is thus a slightly misleading metaphor. Instead of a gradual accretion of knowledge, there are in the sciences various metamorphoses: a continuous transformation and substitution of knowledge often generated from mutually incompatible viewpoints.

(3) *The empirical study of scientific practices.* If rational arguments are insufficient to explain a shift in paradigm, then the task of the philosopher of science requires revision. The rational reconstruction of scientific progress is no longer the question. Furthermore, there is no longer any criterion to determine whether there is progress or not. Scientific theories do not approach *the truth*; they are not even a series of better approximations of the truth. Rather they are incommensurable. Scientists produce theories that

describe the world from dissimilar components. Truth exists at best within the framework of a paradigm; it is a regional enterprise and there is no overriding criterion or rational method to resolve controversy between grossly different paradigms. Verisimilitude is no longer a closure mechanism, a strategy to end a controversy – *power* is usually decisive.

In the 1980s, the social, dynamic, and practical aspects of modern science were radicalized in post-Kuhnian studies of science. Scientific activities are now studied *in situ*. With anthropological and sociological methods, the esoteric culture of the scientific laboratory has been analyzed [12]. Most interesting from this perspective is not ready-made science but science in the making, the examination of scientific controversies [5], i.e., when the production of knowledge is in full process and when a particular view of the world is not yet generally accepted as a "fact."

Thus, a new agenda for the philosophy of science has emerged. Distinctions between internal and external determinants of science and between sociology, history, and the philosophy of science seem of lesser importance nowadays. Today we view scientific facts as constructs; they are, so to speak, negotiable. The biological concept of "growth" applied to knowledge acquisition is better replaced with a concept from engineering: knowledge is effectively *constructed* and *created*. The created products require acceptance and eventual use. Their intrinsic qualities alone cannot do the job. They demand a stimulating context in which to be fabricated and consumed. Science is in fact the building of networks, as Latour argues: "Without the enrollment of many other people, without the subtle tactics that symmetrically adjust human and non-human resources, the rhetoric of sciences is powerless" ([13], p. 145). Cognitive claims become well-established facts through mobilizing a network of allies. Medicine and science are now characterized in Machiavellian terms: "The picture of technoscience ... is that of a weak rhetoric becoming stronger and stronger as time passes, as laboratories get equipped, articles published and new resources brought to bear on harder and harder controversies. Readers, writers and colleagues are forced either to give up, to accept propositions or to dispute them by working their way through the laboratory again" ([13], p. 103). Truth is no longer relevant; facts become true because they are *effective*. One of Latour's illustrations is the work of Pasteur [14]. Pasteur succeeded to mobilize a network of economic, political, and scientific forces in French society to support his own research. Society and its scientific facts are built simultaneously: "... the framework of society was redefined in order to *make room* for the microbes. Elimination of them from the social relations that they distorted also made room for the Pasteurians" ([14], pp. 103-4).

The two strategies analyzed above illustrate how complicated the relationship between medicine and the growth of knowledge has become. On the one hand, modern philosophy of medicine holds that we can no longer assume that medicine practiced with increasing effectiveness is a clear-cut discipline with well-defined boundaries, characteristics, and goals. On the other hand, modern philosophy of science makes clear that the traditional notion – that knowledge is acquired when truth corresponds to reality – must now be discarded. If we accept this new truth, then the growth of medical knowledge must be re-interpreted. There will never be an ideal state of knowledge. Knowledge is the result of scientific communities; the more they succeed in constructing conceptual networks, the greater the claim to knowledge and the more widely a "body of knowledge" will be taken to be true. This is an interesting prospect since in our own time medicine, science, and society are intricately interwoven. Medical scientists and physicians have created very powerful networks that spread throughout various societies and penetrate into every aspect of human life. This new "fact" deserves closer scrutiny. How did medicine's recently-acquired power and effectiveness become established? This volume addresses this central question from various perspectives.

## BIBLIOGRAPHY


1. Arens, W.: 1979, *The Man-Eating Myth: Anthropology and Anthropophagy*, Oxford University Press, Oxford.
2. Boon, L. and de Vries, G. (eds.): 1989, *Wetenschapstheorie. De empirische wending*, Wolters-Noordhoff, Groningen.
3. Brandt, A.M.: 1987, *No Magic Bullet: A Social History of Venereal Disease in the United States since 1880*, Oxford University Press, New York.
4. Butterfield, H.: 1931, *The Whig Interpretation of History*, G. Bell, London.
5. Engelhardt, H.T. and Caplan, A.L. (eds.): 1987, *Scientific Controversies: Case Studies in the Resolution and Closure of Disputes in Science and Technology*, Cambridge University Press, Cambridge, Massachusetts.
6. Forstrom, L.A.: 1977, 'The Scientific Autonomy of Clinical Medicine', *The Journal of Medicine and Philosophy* 2, 8–19.
7. Good, B.J. and Good, M.J.: 1981, 'The Semantics of Medical Discourse', in E. Mendelsohn and Y. Elkana (eds.) *Sciences and Cultures: Sociology of the Sciences* V, D. Reidel Publishing Co., Dordrecht, pp. 177–212.
8. Horstman, K.: 1989, ' De geschiedenis van de medische praktijk: tussen sociale geschiedenis en historische sociologie', *Nederlands Tijdschrift voor Geneeskunde* 133, 781–784.

9. Inlow, W.D.: 1946, 'Medicine: its Nature and Definition', *Bulletin of the History of Medicine* 19, 249–273.
10. Kneucker, A.W.: 1949, *Richtlinien einer Philosophie der Medizin*, Verlag Wilhelm Maudrich, Wien.
11. Kuhn, T.S.: 1970, *The Structure of Scientific Revolutions*, University of Chicago Press, Chicago, Illinois.
12. Latour, B. and Woolgar, S.: 1986, *Laboratory Life: The Construction of Scientific Facts* (1979), Princeton University Press, Princeton, New Jersey.
13. Latour, B.: 1987, *Science in Action*, Open University Press, Milton Keynes.
14. Latour, B.: 1988, *The Pasteurization of France*, Harvard University Press, Cambridge, Massachusetts.
15. McKeown, T.: 1976, *The Role of Medicine: Dream, Mirage, or Nemesis?*, The Nuffield Provincial Hospitals Trust, London.
16. Payer, L.: 1988, *Medicine and Culture: Varieties of Treatment in the United States, England, West Germany and France*, Henry Holt and Company, New York.
17. Sadegh-zadeh, K.: 1980, 'Medizin, Wissenschaftstheoretische Probleme der.', in J. Speck (ed.) *Handbuch wissenschaftstheoretischer Begriffe*, Vandenhoeck & Ruprecht, Göttingen, F.R.G., pp. 406–411.
18. Schaffner, K.F.: 1989, "Introduction" in 'The Structure of Clinical Knowledge', *The Journal of Medicine and Philosophy* 14, 103–107.
19. Shortt, S.E.D.: 1981, 'Clinical Practice and the Social History of Medicine: A Theoretical Accord', *Bulletin of the History of Medicine* 55, 533–542.
20. Szumowski, W.: 1949, 'La Philosophie de la Médecine: son histoire, son essence, sa dénomination et sa définition', *Archives Internationales d'Histoire des Sciences* 2, 1097–1139.
21. Wieland, W.: 1975, *Diagnose: Ueberlegungen zur Medizintheorie*, Walter de Gruyter, Berlin, F.R.G.

SECTION I

# MEDICINE, HISTORY, AND CULTURE

HENK A.M.J. TEN HAVE

# KNOWLEDGE AND PRACTICE IN EUROPEAN MEDICINE: THE CASE OF INFECTIOUS DISEASES

## INTRODUCTION: THE IMPACT OF A STRANGE EPIDEMIC

In early summer 1832, a new disease entered the Netherlands through the port of Scheveningen. During July and August the disease had spread throughout all major towns and cities in a nation-wide epidemic. In fact, the disease entered the Netherlands rather late. It had appeared in Moscow in September, 1830; it invaded the British Isles a year later; it made its dramatic impact on Paris in March, 1832.

This pestilence appeared to emerge from India, and had advanced since 1817 through every country that stood in its path. In many countries, "Asiatic cholera" (as it came to be called) created an atmosphere of crisis as would an imminent foreign invasion. It had been centuries before that Europe had been seriously affected by a widespread and devastating plague. The new pandemic, as might be expected, produced an unprecedented fear. The unpredictable nature and the sudden explosive eruption of cholera, the so-called "Blue Death", revivified the horrors of the medieval and seventeenth-century Black Death, without, however, any empirical justification, since the number of cholera deaths was small compared to earlier plague death rates, and nearly equal to the contemporary death rates of diseases such as typhus and common diarrhoea. For example, in Britain in 1831–32 cholera affected approximately 1 per cent of the population (with 1.6 deaths per 1.000) [27, 32]. More important than the demographic impact were the social and scientific responses to cholera. For the political authorities, the epidemics were an occasion to proclaim and enforce rules and regulations, and to establish boards of health. This governmental action led to a protracted debate on the appropriateness and effectiveness of traditional measures like quarantine regulations, cordons of troops to seal off infected areas, the separation of the sick from the healthy, and to isolate them in their own houses or to relocate them in special cholera hospitals.

No government, liberal or autocratic, could escape from being involved in these epidemics. Cholera was indisputably and generally considered a public health problem. There was a general consensus that society ought to provide for those threatened and affected. In the course of subsequent epidemics (e.g., 1831–2, 1848–9, 1853–4, 1866) a fundamental dispute emerged: whether the



*H.A.M.J. ten Have et al. (eds.), The Growth of Medical Knowledge, 15–40.*

authority and competency to deal with this problem should be vested in the traditional locus of authority (e.g., ruling classes, magistrates, clergy) or the medical profession that claimed scientific knowledge and technical ability? In this dispute the claims of medicine were hampered by the fact that the medical profession was not strongly unified. To argue that authority should vested in those with rigorous practical training and knowledge as well as upon those who had required science-based research (*viz.*, *Lancet,* November 1831) [32], p. 26), presumed that there existed a solid knowledge base for policy decisions. In fact, as I hope to show in what follows, it was not superior knowledge but the actual successes of clinical practices that helped medicine to gain social authority and prestige. Although theories and cognitive frameworks were available to interpret and explain the origin, symptomatology, and transmission of an infectious disease such as cholera, there was, for a long time, no agreement within the profession about the preferential criteria and reliable methods for choosing between competing theories. In this situation only *relative* growth of medical knowledge was possible (*viz.*, within the context of a specific theoretical framework). At the same time, the absence of a comprehensive theory and the lack of rational criteria to give preference to one framework over the other, did not impede the progress of medicine since it evolved through a series of successful practices by means of which it articulated its scientific self-interpretation. The doctors' understanding and management of cholera revealed some aspects of the complex interplay between theory and practice in medicine.

For a number of reasons, the case of the cholera epidemics merits particular attention from the perspective of the philosophy of science. During the 19th century, the philosophical foundations of present-day medicine and health care were firmly established. The constitution of a more or less uniform self-image of modern medicine materialized through a long and laborious process of transformations, debates, and controversies. In this process some crucial issues had to be settled:

(1) How to decide which theory of infectious disease should have priority? Might there be rational criteria to make this decision (such as truth, plausibility or effectiveness), or should preference be given to a specific theory as the result of political power and social support?

(2) How to make theory and practice more congruent? Traditionally, medical practices were very diverse and sometimes incompatible. Although there was an abundance of theories, it was not possible to decide whether this practical diversity was the outcome or the origin of the variety of theories. Until the end of the nineteenth century, a normative methodology was

unavailable. Until then, it was hard to decide *a priori* which theory should guide medical practice, and *a posteriori* which practices were justified.

(3) How to agree on the self-interpretation of medicine as a science? The early controversies concerning the etiology of, therapy for, and prevention of cholera, reflected a fundamental conflict concerning the scientific status of medicine. For some time it appeared that medical practice could successfully deal with cholera if medicine were interpreted as a social science. This interpretation was more and more challenged, however, in the second half of the century.

At the turn of the last century, the outcome of the above noted disputes was clear: medicine was generally held to be a natural science with a powerful paradigm or model to explain and manage infectious diseases. This so-called "infection model" included the theory of specific etiology (particular diseases have particular causes), and the practical maxim of specific treatments for specific diseases. The nature and power of the model was promulgated by public demonstrations and laboratory experiments. The famous historic example was a dramatic performance in May, 1881 (on a farm at Pouilly-le-Fort in France in front of many spectators – politicians, scientists and farmers); Pasteur and his assistants had injected vaccine into sheep and predicted that these animals would survive the administration of a lethal dose of anthrax bacilli.

Because of its precise results as well as its explanatory power, the infection-model turned out to be the most powerful paradigm of modern scientific medicine; it still represents for many philosophers of medicine the core of the biomedical model of disease and treatment. This model has come to define the nature and role of modern medicine.

Focusing on infectious diseases and on the social and scientific responses to them, we can come to understand the theoretical framework of today's medicine. In the last century, medical practice was founded on an ongoing experience with sudden collapse, pain and death, but also on an intense debate between competing frames of reference and conflicting explanatory schemes. To understand the growth of medical knowledge and its impact on medical practice, we need careful and detailed case-histories, studies which not only describe the role of internal factors in the evolution of medicine but also the relationship between scientific and medical advances as well as cultural values and social attitudes.[1] As Michel Foucault has pointed out, this means that one should not write a history of the past in terms of the present, but rather the history of the present ([19], p. 31).

At this juncture I turn to the interactions between medical theory and practice, particularly with regard to infectious diseases, and then sketch two rival theories of disease causation (well-known in pre–19th century medicine) and their influence on the practice of health care. This reveals a growing tendency to bridge the gap between theory and practice by a shift in emphasis in health care from moral reformation toward instrumental and technical innovation. As we shall see, the upsurge of interest in the philosophy of medicine originates in the need to determine the scientific status of medicine. An awakening interest in medical concepts, scientific methods, and a need for a synthesis of clinical experience and basic sciences, resulted in a critical attitude towards the 19th-century notion of medicine as a social science.

## THE TRADITIONAL TEXTURE OF THEORIES

At the time of the arrival of cholera in the West two competing theoretical frameworks were available to explain the etiology, transmission, and prevention of epidemic diseases: contagionism and miasmatism. Both came from a long-standing tradition in medicine. They represent, in the terminology of Ludwik Fleck [15], pre-scientific thought-styles, each in its own way defining both the nature of medical problems and the way to resolve these problems. These styles result from a reinterpretation of ancient 'proto-ideas' which serve as heuristic guidelines regulating research and medical practice.[2] Fleck regards proto-ideas as developmental rudiments of modern theories, originating on a socio-cognitive basis. New knowledge is influenced by these proto-ideas and reflects their original and primitive traits. It is not important whether proto-ideas are true or false; the only important question concerns their *usefulness* as starting-points for the development of scientific facts. The history of science is the evolution of communal *styles of thinking*. Perception, observation, experimentation depend on the prevailing thought-styles. Cognition therefore is a collective activity. Without a community of thought-style no discovery would be possible, no facts would be produced. The product of thinking, according to Fleck, is "a certain picture, which is visible only to anybody who takes part in this social activity, or a thought which is also clear to the members of the collective only" ([16], p. 77).

The arrival of a new and terrifying disease like cholera increased the tension between the prevailing thought-styles produced by the two types of proto-ideas. To set the stage for the 19th-century debate, I shall simply summarize the vital role of these styles in the history of medicine.

### *Contagionism*

Contagionism attributes epidemic disease to minute living organisms. It postulates the conveyance of distinct germs between individuals. These so-called "contagia" reproduce themselves in the individual they have infected, and they pass from the sick to the healthy by direct or indirect contact. The emphasis, therefore, is on the individual person, while the environment is merely a passive medium. From this idea clear directives for medical practice arise: the only effective measure against an epidemic disease is the most stringent isolation of all those afflicted.

The idea of contagion is first found in the Bible, particularly in *Leviticus*. In ancient Greek and Roman medicine, contagionist ideas were not very influential, but their importance grew after the sixth century with the spread of leprosy over Europe. Medical health regulations were inspired by contagionism: quarantaine laws, pesthouses and leperhouses were intended to interrupt the dissemination of disease. Towns, provinces, and countries were sealed off by a cordon of troops in an attempt to keep out pestilence. The infected individual as carrier of contagia was the prime target of a policy of isolation. The idea of contagion was never so much emphasized as between 1350 and 1500 when the plague decimated the population of Europe.

Contagionism was formulated as a coherent, but speculative theory by the Renaissance scholar Fracastorius (1478–1553). In his book *De Contagione*, published in 1546 [20], he presented a theory of infection, a classification of contagious diseases, and prescriptions for their treatment.[3] Fracastoro postulated the existence of small imperceptible particles as causes of disease *(seminaria morbi)*. Admitting that it seemed odd to suppose that infection is caused by invisible particles, he tried to make his thesis theoretically acceptable, using arguments by analogy.

Common sense observations such as the rotting of fruit in a basket and passing to adjacent pieces, were usually explained by the *exhalation theory*. Every substance is continuously releasing particles; these emanations and evaporations are detectable with our senses (particularly the nose). The disease-causing particles are therefore not unique according to the exhalation theory but they differ from the other ubiquitary particles by their negative property of transmitting infection [22]. From the plausibility of the existence of *seminaria morbi* Fracastorius proceeds to explain how these invisible particles passed from individual to individual. He distinguished three modes of infection: "*contactu*", "*per fomitem*"[4] and "*ad distans*". The quality of the particles determines its mode of infection: if they are 'tough' and 'vigorous' they adhere to objects like wood or clothes, if they are 'weak' and 'unstable' they only infect through direct contact.

Within the framework of humoral pathology Fracostorius further demonstrated how different types of particles have a specific affinity with particular components of the body and how they produce pathological effects. The specificity of the *seminaria* is explained with an appeal to the more general principle of "sympathy": objects only interact with particular others which have a similar constitution. The pathological effects are also explained within the current framework of Galenism. An imbalance of the body fluids resulting from obstruction, plenitude, or corruption, produces putrefaction. This process is generating *seminaria morbi*. As soon as they exist, they have the power to propagate and engender themselves; they spread to healthy individuals and disrupt their humoral balance.

Fracastorius' theory of communicable diseases had considerable impact on his contemporaries and on the following generations of physicians.[5] During the 17th century the development of the microscope provided a new impetus to contagionism. An important contribution was made by Athanasius Kircher (1602–1680), who in 1658 published the data of experiments with a microscope to find the postulated micro-organisms which caused disease. He claimed to have identified 'tiny worms', minute animals, as the causative agents of infection [29]. Contrary to Fracastorius, who never specified that *seminaria* were living organisms, Kircher made clear that his worms must be considered as animate contagia.

His animalcular theory which explained the behaviour of epidemics by actually visible, living and multiplying organisms, aroused much interest. Between 1650 and 1800 many physicians in European countries wrote on contagionism and on their own observations and experiments with assumed germs of all kinds of diseases. Particularly between 1700 and 1730, meticulous studies of disease behavior and an improved use of microscopes, initiated a thought-style with a high degree of medical sophistication, anticipating notions and ideas developed later in bacteriology [41]. In the same years the practice of inoculation against smallpox was beginning to attract public attention; the success of this empirical technique could easily be explained from the viewpoint of the animalcular theory.

After the 1730s, interest in the contagion idea declined. Although no new experimental results or original contributions were published, contagionism remained alive in the minds of many physicians. A few authors kept the idea alive without providing new confirmation. On the basis of the methods and techniques available in those days, it seemed impossible to decide the issue between the contagionists and their opponents.

In the second half of the next century, however, the animalcular views of the contagionists were revived and became the basis of a new scientific approach, which led to the establishment of modern bacteriology.

### *Miasmatism*

The second proto-idea, in constant controversy with contagionism, can also be traced to Antiquity. It is the idea that disease is produced by odours and emanations, i.e., drifting unhealthy miasmas. These miasmas are generated by a variety of sources in the human environment: dirt, bad drainage, crowded housing, lack of ventilation and polluted water. Generated under specific conditions of filth, temperature, and moisture, the poisonous miasmas are propagated through the air, not through patients. Disease is not an individual affair but a product of the *environment*. The presence of miasmas could be detected by smelling. In summary: not contagion from person to person is responsible for eruptions of disease, but rather a widely diffused, atmospheric agent.

The miasmatic theory of disease causation was first stated in the Hippocratic writings, particularly the treatise *Airs, Waters, Places* [25]. The morbid influence of the environmental conditions was not explained in a detailed theory of the association between disease and environment, but a number of specific ideas about disease-promoting circumstances were introduced. Climatic conditions and atmospheric changes may cause epidemics. For example: the alterations in a season determine the character of diseases; south winds cause deafness, dimness of vision, headaches, whereas a north wind causes coughs, sore throats and constipation. Knowledge of these environmental complexes is necessary for the successful practice of medicine.

The author of *Airs, Waters, Places* issues his creed just at the start of his treatise: "Whoever wishes to pursue the science of medicine, must proceed thus. First he ought to consider what effects each season of the year can produce.... The next point is the hot winds and the cold.... He must also consider the properties of the waters" ([25], p. 71).

When a physician arrives at a town with which he is unfamiliar he should examine its site, elevation, soil, climate, winds, drinking waters and diet. Doing so, he could predict which diseases would prevail, and he would know which treatments were advised.

In the Hippocratic tradition humanity is regarded as a microcosmos, mirroring the constellation of the macrocosmos. Because of this interrelation, disease is not an isolated event, not the result of a disorder within mankind, but an integral part of nature, the product of disorder between mankind and environment. Diseases actually are symptoms of the seasons.

After the pinnacle of Hippocratic medicine, miasmatism fell into disregard. Drawing little attention in the Middle Ages, it reached a new peak of acceptance and elaboration in the eighteenth century. That the idea of a miasmatic origin of epidemic disease was not completely forgotten, is found in three reasons:

1. In Greek and Roman Antiquity malaria was commonly accepted as the paradigm case of a miasmatic disease.

   The association between this particular ailment and stagnant water did not escape the physician's notice. Nobody seriously doubted that swamp emanations or 'marsh miasma' produced malaria.
2. The lasting influence of Galen's three-fold classification of factors affecting health: the naturals, the nonnaturals and the praeternaturals. Health is considered to result from an absence of praeternaturals (things anti-nature, disease), from a proper ordering of the naturals (the innate structural and functional components of the human body) and from a proper regimen of the nonnaturals. This last category is the most interesting.[6] In the doctrine of Galenism six nonnaturals are mentioned: ambient air, food and drink, sleep and watch, motion and rest, evacuation and repletion, and passions of the mind. These factors are extrinsic to the natural constituents of the body but necessary for the preservation of health. The physician's task is to regulate the way in which the patient handles and is affected by these six sets of factors. For centuries an important aspect of therapeutics has been the regulation of the nonnaturals: medicine is seeking to order the routines of daily living by means of a dietetic regime [13]. The nonnaturals focus attention on the intimate connection between living conditions and personal free choice.[7]
3. The attention paid in medicine to the quality of the surrounding air. In Greek, Roman, and Islamic medicine it was common knowledge that health is best preserved by breathing pure air, not polluted by vapors from marshes, pools, swamps, or sewers.

   In the 17th century, the idea that impurity of air is one of the causes of disease was endorsed by Thomas Sydenham (1624–1689), "the English Hippocrates". Urging the importance of clinical observation and case histories for a proper theoretical understanding of disease, Sydenham stressed that a disease finds its origin in man's habitat. By meteorological observations and intensive study of environmental variables as well as mortality records, Sydenham could specify various "epidemic constitutions" of the atmosphere. Certain epidemic fevers apparently are produced by a distinct combination of atmospheric and environmental phenomena. Inhalation of elements from this constitution brings about disease.

Construing the atmosphere as a reservoir of disease, attempts are made to purify the air from noxious emanations. Hippocrates and Galen advised their patients against living in the neighbourhood of swamps and stagnant water. Celsus advocated that a patient with fever should be kept in a room where he can inhale plenty of pure air.
During the Black Death in the 14th century the corrupted atmosphere was purified by burning dry and richly scented woods or barrels of pitch and tar in the streets, and by filling the houses with pleasant smelling plants and flowers, and sprinkling the floors with vinegar and rosewater. On the whole, the attitude towards environmental phenomena was quite fatalistic and passive. The environment was regarded as an unchangeable matter of fact, and the atmospheric causes of disease as beyond human control. Site and climate, soil and air seemed unalterable.

In the eighteenth century this pessimism and fatalism gave way to a new active approach. We witness the emergence of a new thought-style, resulting from *the specification of old miasmatism into a new environmentalist theory*. This theory starts from the idea that environmental conditions might themselves be modified.

The environment must be put under surveillance; in this way the epidemic cycle could be interrupted either by removing man from the miasma-ridden environment or by cleansing this environment. An outstanding representative of this "environmentalism" ([37], p. xv), this "medicine of climates and places" ([18], p. 51) was John Pringle (1707–1782). He was a student of the famous Dutch physician Boerhaave, and became professor of Pneumatics and Moral Philosophy at Edinburgh, and in the 1740s physician of the British army in Flanders. As an army physician Pringle started a series of investigations to detect an association between environmental conditions and diseases: "I inquire into the more general causes [of disease] ... namely such as depend upon the air, the diet, and other circumstances usually comprehended upon the appellation of the non-naturals" ([35], p. vii). Of all the nonnaturals, noxious and putrid air was considered the most important cause of sickness. The Low Countries in particular proved to be unhealthy because of the evaporation of the polluted water in the numerous canals and ditches as well as the stagnation of the air in that part of Europe. Armies were ridden with infectious diseases because a multitude of men in a relatively small area produced effluvia changing the air into a fatal medium.

Pringle also studied the nature of hospital and jail-fevers, and he made the same observations. Both institutions are frequently exposed to epidemic diseases, as they are ill-aired and in a constant state of filth and impurity. To explain his observations, Pringle developed a detailed theory of the environmental forces underlying disease. He gave an exact specification of the various sources of miasma. The miasma, consisting of the effluvia from putrid substances, was the external cause of epidemic disease and was omnipresent in society: "From this view of the causes ... it is easy to conceive how incident they must be, not only to all marshy countries after hot seasons, but to all populous cities, low and ill-aired; unprovided with common sewers; or where the streets are narrow and foul; or the houses dirty; where fresh water is scarce; where jails, or hospitals are crowded, and not ventilated or kept clean; when in sickly times, the burials are within the towns, and the bodies not laid deep; when slaughterhouses are within the walls; or when dead animals and offal are left to rot in the kennels, or on dunghills; when drains are not provided to carry off any large body of stagnating or corrupted water in the neighbourhood... " ([35], p. 324).

Pringle departs from convention in recommending specific measures to be taken to prevent disease. Ventilation or mechanical airing to dispel the noxious air is most important. When that cannot be done, purification of the streets, houses or wards is necessary, using chemical means to reduce stench, for example, diffusing the steams of vinegar or burning tar, tobacco, even gunpowder. Giving medicines is useless; there can be little hope of recovery while the corruption of the air continues. The idea that epidemic diseases arise from specific factors in the environment had an increasing number of advocates in the eighteenth century. Pringle simply exemplifies the growing sympathy for miasmatism.

Eighteenth-century environmentalism brought about two essential modifications in miasmatism which might explain the growth in popularity of miasmatic thinking into the 19th century:

(1) The claims of miasmatism had become more specific and were

stated with less dogmatism. The contagious nature of a limited number of diseases (e.g., syphilis and smallpox) was acknowledged. This means that miasmatism no longer implied an absolute anti-contagionism; it no longer needed to develop a complicated theory to explain contradictory observations. Accepting that the theory of contagion offers a plausible explanation of the etiology of at least *some* diseases, it introduced a general framework of two different types of communicable disease. At the same time, miasmatism maintained with double confidence that the most important contemporary health problems were best explained and avoided by assuming a miasmatic origin. This readjustement increased the acceptability of the miasma theory. In fact, from then on a distinction could be made between *epidemic* and *contagious* diseases.

(2) The notion of miasma was specified as a set of variables that rendered miasmatism accessible to empirical research. At first miasmatism was a rather vague theory, designating uncontrollable conditions as sources of disease. In the eighteenth century, with the transformation of miasmatism into an environmentalist theory the nature and origin of miasma was specified. Instead of atmospheric constitutions and ill-defined emanations, specific factors such as meteorological changes and wind-directions, the construction of dwellings, stagnant waters, the heaps of refuse and rotting waste matter in the cities, were blamed and subsequently attracted research. By compiling epidemic histories, meteorological journals, mortality registers, and topographical data, physicians such as Locke, Petty, Sydenham, and Frank tried to gather empirical evidence to document an association between environmental factors and diseases [37]. These investigations were guided by the idea that epidemic disease itself could be avoided by manipulating the miasma producing environment.

## THE 19TH-CENTURY CONTROVERSY

The arrival of the cholera in Europe initiated a new phase in the controversy between contagionism and miasmatism. In the 1830s, contagionism was the common-sense view; it was the favorite doctrine of the medical establishment and it had come to dominate official policy. In England, for instance, the Privy Council, the Central Board of Health, the Royal College of Physicians, and the recently-established Lancet all declared, in 1831, that cholera was a communicable disease, transmitted by minute self-reproducing creatures from one person to another ([32], pp. 23 ff). This preference for contagionism implied that the greatest importance was attached to isolation: quarantining ships, isolating diseased towns and villages, and, finally, isolating the victim himself. Many governments in Europe felt obliged to take rigorous precautions. These involved a considerable element of coercion which led in some cases to popular violence and resistance; in less autocratic societies, more voluntary measures were taken. But whether coercive or voluntary, policies in various countries followed the same basic pattern: quarantine and cordon sanitaire, setting up cholera-hospitals to isolate the sick from the rest of the population, purification of houses, destruction of personal belongings, and immediate interment in special burial grounds outside urban areas.

However, the fact that despite these measures cholera could not be stopped, discredited contagionism. Contagionist explanations were being challenged with increasing success. During the first epidemic even supporters of contagionism noticed that cholera adhered to none of the theoretical rules: the disease was advancing more rapidly than could be explained by transmission from individual to individual; quarantine measures were less than effective; the disease did not spread in a continuous manner; in an affected country whole areas were spared; even streets and some houses might stand entirely untouched; cholera seemed to prefer the most densely populated and filthy areas.

These observations of cholera's eccentric behavior and curious geography constituted anomalies for contagionist analysis, yet they were not decisive in ending the debate. Contagionists could show that cholera followed lines of human communication, and that in many cases contact with infected places could be traced.

The choice between the two thought-styles, however, was only partly a matter of argumentation and rationality. Scientific standards gave no clear guidance. How difficult it was to resolve the debate was demonstrated in a letter by the British physician Walker, writing from Moscow where he witnessed the appearance of cholera in September, 1830. He wrote to his government: "I intended to say that I myself am *convinced* of the contagious nature of the disease, but that the *proofs* of its transmission from one individual to another are not quite perfect as yet" ([7], p. 111). These remarks illustrate the tension brought about by the actual confrontation with cholera, between the dominant belief-system of contagionism and the scientific requirements of objective observation and evidence.

As the epidemic spread, physicians who tended to be contagionist were converted to the miasmatic view. At some places the doctors settled the question democratically by a show of hands; for example the Westminster Medical Society voted in 1832, by a narrow majority, that cholera was not contagious ([32], p. 70). The controversy split the medical profession; the two parties constantly attacked each other in print as well as at public meetings. The French pioneer of physiology and experimental medicine, François Magendie, ridiculed the idea that specific diseases were caused by specific germs. The Dutch general practitioner, J. d'Aumerie, writing his memoirs of the cholera at Scheveningen, mocked the notion that little animals were infecting their fellow-animal: "It will not be necessary to falsify these dreams. I only put them forward to demonstrate into what foolishness even wise men could lapse" ([10], p. 125).

Doubts and scepticism toward contagionism made the theory of a miasmatic origin look more attractive, especially because in the preceding period it had undergone essential modifications. After the first cholera epidemic contagionism had lost its dominance. The idea that a healthy environment was indispensable for the prevention of disease was taking root among the more enlightened members of the middle classes and among the majority of the medical profession. During the second pandemic (1848–49), miasmatic thinking dominated medical research and public policy, whereas during the third epidemic (1853–55) official and professional opinion was swinging back to contagionism. Ten years later, with the fourth cholera pandemic (1865–67), the majority of the profession again accepted a contagionist theory.

The oscillation of the influence of contagionism and miasmatism is a remarkable phenomenon in the history of medicine. It is amply documented in Great-Britain [7, 29]. It is illustrated by the changing opinion of Sir John Simon, the first Medical Officer of Health for the City of London, who was an enthusiastic miasmatist in 1849 and an equally convinced contagionist in 1865 ([42], pp. 287 ff). In the United States, miasmatism was already the dominant theory when the cholera arrived. But the number of supporters of a contagionist etiology slowly rose until they finally outnumbered the miasmatists in 1866 [39]. These changing preferences point to the fact that the majority of European physicians believed in a contagionist etiology of cholera years before the major bacteriological discoveries of the 1870s and 1880s, especially the discovery of the Vibrio Cholerae by Robert Koch.

The dominating influence of miasmatism was waning in the same period when Pasteur experimentally refuted the concept of spontaneous generation and formulated his germ theory of disease, and when Lister reported on the success of antiseptic surgery.

The achievements of contagionism in the last decades of the 19th century, which are usually considered as a triumph of scientific medicine do not, however, detract from the fact that the apparently false theory of miasmatism had a positive effect on public health, and actually contributed to a decline in European mortality rates since the eighteenth century.[8] The loss of appeal which befell to miasmatic theories about the middle of the last century should therefore not be attributed to their lack of applicability or efficacy in medical practice. On the contrary, it seems to have been the result of a complicated process in which the nature of medicine itself was redefined, that is, the object of medical practice was reformulated.

The fluctuation in popularity of contagionism and miasmatism illustrates the persistent tension between individualistic and social viewpoints in health care as well as a conflict between self-interpretations of medicine as a natural or social science. The continuity of these proto-ideas is also an indication of the impossibility of resolving such controversy with empirical data or scientific arguments alone. In fact, the controversy was ended towards the close of the nineteenth century. At that time, the new natural scientific approach prevailed and brought to an end a long succession of alternating but related thought-styles. This phenomenon deserves some elaboration.

## GROWTH OF MEDICAL KNOWLEDGE

The arrival of cholera re-activated the latent conflict between contagionism and miasmatism. The peaceful coexistence of two explanatory frameworks, each with its own set of prototype diseases (e.g., syphilis versus malaria) was suddenly challenged by an entirely unknown and fatal pestilence.

How did this interaction promote the growth of medical knowledge? The answer is complicated, since in both frameworks different types of knowledge seemed to have priority.

In miasmatism, the concept of miasma was not analysed or developed further into a more sophisticated theory. The theory as such did not attract scientific attention, but rather the perfection of its application. Within the miasmatic framework there was primarily a growth of instrumental, practical knowledge concerning the sanitation of the environment. This pragmatic attitude prevailed, for example, in the works of Edwin Chadwick, the moving force behind the British sanitarian movement. In his opinion, precise knowledge of the *causes* of a disease is irrelevant for its cure and prevention. In the past, leprosy, plague and the mysterious sweating sickness, all disappeared without any theoretical knowledge of their causes. We need to know how effectively to control and modify the *environmental* conditions that facilitate the spread of disease. Scientific disputes on the origin of disease will only hamper the growth of medical knowledge: "The medical controversy as to the causes of fever ... does not appear to be one that for practical purposes need be considered, except that its effect is prejudicial in diverting attention from the practical means of prevention" ([9], p. 214).

For miasmatists, medical knowledge was not of interest in virtue of its truth, but because it worked. And the curious fact is that practical knowledge was indeed very effective. In contagionism, on the other hand, the knowledge sought for was primarily (and for a long period) *theoretical* knowledge. The main problem in contagionist thinking was the methodological proof that contagia were not technical artifacts or the products of disease but their direct causes. Even this problem required some practical knowledge, of course, namely, how to develop microscopic techniques and staining methods necessary for the objective identification of contagia. But this practical knowledge was only instrumental regarding the theoretical goal of irrefutably establishing that contagia or *germs* were the causes of particular diseases. Contagionists' theoretical knowledge really increased with the formulation of Koch's postulates in the 1880s.

Although contagionism as well as miasmatism were associated with a growing body of knowledge (theoretical and practical, respectively), neither in theory nor in practice were the successes decisive in changing the popularity of both thought-styles. In the 1840s, miasmatism became popular before its practical effects had been demonstrated, whereas contagionism again became dominant from the 1860s, despite the fact that it had not yet evolved into a scientific bacteriological theory and despite the effectiveness of medical practices based upon miasmatic thinking to improve the social conditions underlying the prevalence of epidemic disease.

Cognitive factors or practical results apparently did not decisively determine the acceptability or plausibility of a theory for the medical profession. The end-result of the long-standing controversy between miasmatism and contagionism was the undeniable growth of medical knowledge. But this growth, laying the foundation of today's medical power, resulted from the changing of the style of thought. Medical science was developing mostly through the dynamic interaction of thought-styles. These thought-styles, argued Fleck [15], are collective phenomena. They point to the fact that knowledge is socially conditioned. Miasmatists and contagionists belong to social networks. As miasmatism and contagionism are embedded in medical and cultural history, they are also embedded in the social context.

The oscillation between contagionist and miasmatist thought-styles in the 1840s can be explained by changes in this context. The initial response to the cholera during the first epidemic was dominated by contagionism. It was the thought-style of highly educated physicians and state authorities. However, the measures taken implied loss of trade and disruption of family life. Quarantine was not based upon any knowledge of the disease but upon tradition and authoritarian power ([32], p. 30). Thus miasmatist opponents in medicine could form an alliance with commercial interests and political liberals to attack the dominant contagionist thought-style. After the first cholera epidemic it was clear that the contagionist policy had failed. Contagionism was associated with a bureaucratic and coercive strategy of external control, *endangering individual freedom* with what turned out to be ineffective measures. In fact, this strategy was based upon the concept of medical police: a programme for state action in the sphere of health and welfare. It arose from the political, economic, and social basis of the autocratic German states in the late eighteenth century [1, 38]. What is unusual is the range of the proposals, the aim of the program, and the methods for achieving it.

The first point is illustrated by the proposals made by Franz Anton Mai (1724–1814), professor of obstetrics at Heidelberg University. In 1800, he published a draft of a health code. The topics covered indicate the comprehensiveness of this proposal; everthing is apparently related with health: from procreation, nutrition, housing, clothing and recreation to occupation, accident prevention, and education.

The purpose of medical policing is first of all to ensure and enhance the power of the state. Health promotion is not an aim in itself, but only a means to a political end. This is evident from theoretical expositions of the police concept. Well-known is the definition of Von Justi: "The aim of policing is to make everything that composes the state serve to strengthen and increase its power, and likewise serve the public welfare" ([12], p. 7). The health of the people is a matter of direct political concern. The state has an obligation to protect and assure the health and welfare of its members. Because of the vital importance of the aim, the methods for achieving it are authoritarian and paternalistic. If the state has to care for the health of its citizens, coercion, even brute force to protect or restore health, could be necessary.

Although its effectiveness was still unproved, miasmatism became more and more attractive to social reformers and advocates of economic and political liberalism. The miasmatic thought-style can be considered a notable specimen of the 19th-century New Enlightenment, combining a preference for science and a preoccupation with social problems and social activism with a profound sense of change and development of reality and social order ([3], pp. 302 ff). Belief in progress was most characteristic of this new spirit of enlightenment; the world is changing for the better, because, Mill [31] said, most of the great positive evils of the world are removable.

In England for instance, miasmatism served as the nucleus of a growing and successful network of allies: physicians, lawyers, politicians, engineers, and philosophers. This sanitary movement was deeply influenced by the *utilitarian philosophy* of Jeremy Bentham [22]. Leading men like Edwin Chadwick and Thomas Southwood Smith were his disciples. Bentham defined medicine as "hygiastics", *viz.*, "the art of preserving as well as restoring Health" ([5], vol. 8, p. 8). Health is nothing more than the non-existence of disease. Bentham's influence upon the sanitary movement can be summarized as follows:

a) He re-emphasized the social origin of disease. As the conditions of the perfectibility of man are of external nature (for example, legal sanctions), man is also corrupted and diseased by environmental causes.[9]

b) He stipulated that these environmental conditions can be investigated with the latest methods of science, in particular statistics. Empirical description and classification of social problems as well as quantification of data are the characteristics of the *Report on the Sanitary Condition of the Labouring Population of Great Britain*, published in 1842 by Edwin Chadwick [9], one-time secretary of Bentham.

c) Prevention comes first, according to both Bentham and Chadwick. This priority is not a matter of principle but it is profitable and beneficial to the greatest number of the population; it is therefore a typically utilitarian approach. The construction of sanitary facilities, it was argued, has a two-fold preventive effect: economic and moralistic. In the long run, these facilities will produce financial benefits by reducing expenses for curative health care and poor relief. But sanitation of the physical environment will also lead to changes in lifestyle and morals of the lower classes. It was Chadwick's view that, for example, an adequate water-supply will breed hygienic habits.[10]

In England, miasmatism could be transformed into an effective programme of social reform, advocated by an alliance of scientists, technicians and politicians, and philosophically underpinned by the utilitarians.

## MEDICINE AS A SOCIAL SCIENCE

How do we explain the second transition (*viz.*, that from miasmatism to contagionism in the 1860s and 1870s) which ultimately promoted the growth of medicine into a modern science? The answer must be sought in the *utility* of miasmatism in everyday practice which was so conspicuous but upon further reflection rather paradoxical.

The successful 19th-century application of the miasmatic thought-style implied that medicine was also a social science. It was in fact this particular concept that was discarded in the subsequent development of medicine. Why did that happen? What were the reasons that the idea of an association between man's habitat and infectious disease came to be regarded as an obstacle to the growth of medical knowledge? To clarify these questions one must study the intrinsic characteristics of the thought-style which evolved from the miasma proto-idea in the early part of the 19th century. This new style arose from a further specification of eighteenth-century environmentalism, concentrating upon specific social conditions within a broad environmental complex of causes of disease. It was developed simultaneously through several thought collectives: French "idéologues" (Hallé, Villermé), English sanitarians (Chadwick, Southwood Smith) and German liberals (Virchow, Neumann).

Although each of these collectives has its own specific character, they all share two central presuppositions:

(1) The predominant determinants of health and disease are *social conditions and personal behavior*, not the structure and function of the body. For Virchow, disease was an expression of individual life under unfavorable conditions. The normal and abnormal social circumstances under which people live must be subjected to scientific investigation; they are the proper object of medicine.

(2) The nature of medicine is primarily *political and educational.* Physicians are the natural attorneys of the poor; they are best qualified to discover solutions for society's basic problems. "For if medicine is really to accomplish its great task", Virchow declared in 1849, "it must intervene in political and social life" ([22], p. 307).

Both presuppositions were naturally absorbed by the thought-collectives, as the expression of a new and powerful thought-style, i.e., *hygienism.* This style emerged from a fusion of the older concept of medical police and contemporary rationalistic philanthropy.

In the liberal climate of early 19th-century France and England, the idea of medical police became more and more unacceptable because of its absolutist framework. The concept of policing, however, gradually obtained a new, non-repressive sense because it was annexed to an encompassing philanthropic strategy to rationalize social assistance. For example, in Holland, the Society for Public Utility (Maatschappij tot Nut van 't Algemeen), founded in 1784, played a prominent part in organizing philanthropic activities, and attempted to replace charity by effective advice. In England, Benthamites rejected charity as well as the allowance system for their irrationality and lack of utility. "In this country", as Jeremy Bentham summarized his criticism, "under the existing poor laws, every man has a right to be maintained, in the character of a pauper, at the public charge; under which right he is in fact, with a very few exceptions, maintained in idleness" ([5], vol. 8, p. 401).

Neither repression nor public and private charity were considered rational solutions to the problems posed by indigence and disease. As enlightened men, the philanthropists coupled assistance with moralization. Their mission was to launch a large scale offensive to civilize the habits and life-styles of the masses. Philanthropic interventions, as a Dutch author stated in 1851, aimed at creating a man who could be master of his own body, who could control his passions and habits [11]. Norms of behavior, such as cleanliness, economy, domestic nursing, soberness were transmitted not by repression or coercion but by the subtle means of advice, persuasion and education. Material aid was the vehicle of moral influence. The basic mechanism of moralization was that of exchange: if children were properly cared for, families could send them to school; if workers deposited a portion of their resources in saving accounts, they would achieve greater family autonomy. Assistance was a kind of investment; the guarantees must be in the moral rectitude of its recipients. This implies that granting assistance was made conditional on a continuous social surveillance, i.e., "a painstaking investigation of needs by delving into the life of the poor recipient" ([12], p. 68).

The net result of such a system of supervised freedom was the normalization of individual behavior: the norms of a healthy, regular, and disciplined conduct passed into domestic life. The object of assistance is also the object of correction and normalization.

Hygienism thus embodies a series of social technologies to produce a new behavior pattern in specific groups of the population; they are not set up to suppress and punish undesirable behavior but to supervise it, to steer it into other channels.

They function by applying a system of tutelage and contract, described by the French sociologist Jacques Donzelot [12]. If individuals comply with defined standards they are allowed a certain autonomy and responsibility for their own behavior. Positive integration of the majority of the population is accomplished through an informal contract. If individuals escape from this normalization process by not complying with the contract or when small groups with a common interest are formed, which complicate normalization or make it impossible, autonomy will be taken away and replaced by tutelage.

A negative integration is accomplished through a chain of interventions adapted to various deviations from the norm. This second level of regulatory and corrective techniques has to be visibly present to enable the first level to function. One of the incentives to accept the normative standards is the threat to have one's autonomy taken away – the risk of tutelage.

This system of contract and tutelage joined sanitary and educative objectives with methods of economic and moral surveillance. It was established by setting up connections between the medical domain and the social domain. Under the thought-style of hygienism, the formerly separate activities of assistance and repression were brought together.

The annexation of the philanthropic strategy initially enhanced the effectiveness of the hygienic approach. But at the same time it has created a tension within hygienism itself, which eventually led to its decline. Philanthropists and physicians agreed that social and moral deterioration were interconnected. But because of their miasmatic tradition physicians and sanitarians tended to accentuate clean and healthy social circumstances as a prerequisite for the reformation of private morality. Advocating norms of cleanliness for example is only effective after the construction of water-supplies, as Chadwick pointed out. An unhygienic lifestyle therefore is considered a symptom of an unhealthy social order. For philanthropists the relation between personal lifestyle and social circumstances was just the reverse. However, the alliance of philanthropy and miasmatism produced an effective combination of social and moral viewpoints, operating with the system of contract and tutelage. Both allies sought to organize virtue; both tried to introduce the same middle-class norms like cleanliness, industry, temperance, tenderness, although by somewhat different means. The crucial point is that they exchanged the coercive external control of contagionism for a less visible system of internal control. They attempted to model the pattern of life by having norms internalized. Doing so, they in fact brought about a more penetrating control and power over individuals than their contagionist opponents.

The same alliance was also responsible for the decline, or rather the transformation of hygienism in the last decades of the 19th century. Because of its intrinsic emphasis on moralization and normalization hygienism alienated an ever-increasing number of modern physicians; physicians, who like the hygienists were committed to the idea of a radical reform in medicine, in the end came to look upon hygienism as a major impediment to their objects. Originally both parties were aspiring to renovate medical thought and practice by using the concepts and methods of modern sciences such as chemistry, physiology and pathology. In the Netherlands this coalition was successful in establishing a professional organization (1849, "Nederlandsche Maatschappij tot bevordering der Geneeskunst") and a leading medical journal (1857, *Nederlands Tijdschrift voor Geneeskunde*) and in lobbying for the reform of medical education which was realized in 1865 ([24]). But after these successes it came to a rupture between the two medical parties. The propagation of social norms on the part of the hygienists compromised the recently built-up image of medicine as a value-free, objective science. Modern reformed medicine is a practice of expertise, a technology of managing diseases of individuals, that does not impose anything – neither new nor old norms – and is therefore anxious to avoid political issues. It is indeed a paradox that the eventual elimination of hygienism, and the discrediting (for the time being) of miasmatism in general, could succeed just because of their positive results in practice. The strategy of moralizing and civilizing the population is crowned with success; hygienic norms, for example, had been accepted and internalized. Imposing norms of health seemed no longer necessary since these were maintained by autonomous individuals themselves. Hygienism as a normative science, with the explicit intention to spread specific norms among the population, was ultimately superseded by laboratory medicine applying the contagionist discoveries in the basic sciences and employing more subtle, implicit transactions of norms. This was the result of the growth of medical knowledge through mutually interactive thought-styles but it also underlay the later progress of modern medicine.

## SCIENTIFIC MEDICINE

A central thesis in the epistemology of Gaston Bachelard [2] is that the problem of the growth of science must be formulated in terms of *obstacles*.

There is no history of science without shadows, without failures, dissensus, and conflicts. Likewise, the evolution of modern medicine must not be considered only a process of construction but rather a process of deconstruction of knowledge as well. Similar ideas were put forth by Fleck at almost the same time.

The success of medicine, then, in conquering epidemic diseases cannot be explained as a continuous accumulation of knowledge nor as a linear progression of science. On the contrary, modern scientific medicine is the result of rupture and discontinuity, the product of controversial interactions between two different styles of thought. Each style could itself present a rational reconstruction of the evolution of medicine from its own perspective. Within its conceptual framework the quantitative advancement of medical science is indeed possible. But as soon as a new and fatal disease like cholera arrived, it made physicians perceive the epistemological obstacles, and a real prospect of qualitative growth of medicine in case of infectious diseases had opened up. The conflict between contagionism and miasmatism led to a long struggle utilizing all kinds of ammunition and reflects a constant flux among allies.

The outcome of this struggle, i.e., modern medicine with its influential "infection model", illustrates George Rosen's remark: "Whether and how scientific and medical knowledge is brought to bear on health problems not infrequently depends more on the interests and ideology of politically and economically powerful groups than on medical or scientific validity" ([38], p. 2). It is also illustrative of the impossibility of making a clear distinction between internal and external factors in the growth of medical knowledge. What is *medical* or *social* is not objectively demarcated. Modern medicine and its assumptions regarding reality, scientific facts, health and disease, as well as its self-interpretation of what is really "medical", are incomprehensible without an appreciation of its historical and cultural contexts. Medicine, in short, is only one specific manifestation of history and culture.

*University of Limburg,*
*Maastricht, The Netherlands*

ACKNOWLEDGEMENT

I am grateful to my colleagues, Stuart F. Spicker, Paul J. Thung, and Helen Keasberry, for their criticism and suggestions.

## NOTES

[1] Careful and detailed case-histories are, for example, recent studies on cholera [14, 32, 34] and on venereal diseases [6]. These books are answers to Richard Shryock's earlier call for an integration of external and internal factors in medical historiography [40].
[2] Fleck pointed out that "... the prehistory of thought presented to posterity some guiding principles for further development of thinking ... " ([17], p. 96). One of his examples is the proto-idea of *animalia minuta* as a cause of infectious diseases ([15], p. 36; [17], p. 95).
[3] Fracastorius is well-known in the history of medicine; with the publication of his poem *Syphilis sive morbus Gallicus* (in 1539) he introduced the appellation "syphilis" for the venereal disease that affected Europe from 1494 onward. The poem was translated into several languages and exerted an important influence on medical and public opinion. See [43].
[4] Particles can be transmitted from individual to individual indirectly through "fomites", i.e., objects which hide the *seminaria* within pores in their substance. Thus a "fomes" in modern terminology is the nidus or focus of infection.
[5] It is remarkable that the ideas of Fracastorius are either overvalued or undervalued. It seems preposterous to say that his work "... contains the first scientific statement of the true nature of contagion, of infection, of disease germs and the modes of transmission of infectious diseases" ([21], p. 500). It is also presumptuous to denounce his ideas as "rigmarole" ([26], p. 63). Both valuations start from the perspective of modern bacteriology. Compared to present knowledge, Fracastorius' ideas may seem ridiculous, but we should not forget that he tried to develop a theoretical explanation of scientifically unknown and socially important phenomena. Within the context of contemporary science and philosophy his explanation was completely rational; as such it promoted a scientific approach of infectious diseases, facilitating the much later growth of modern bacteriology [22].
[6] For recent literature, see [8, 28, 33, 36].
[7] At the close of the nineteenth-century this traditional component of medical theory had disappeared altogether: the nonnaturals are no longer mentioned as relevant predisposing causes of health. The rise of bacteriology and cellular pathology in the later decades of the nineteenth-century had been accompanied by an increasing focus of all attention on direct causes and mechanisms of disease and a loss of interest in wider aspects of health and disease. However, recent interest in health sciences, particularly health education and health policy, seems to revive the traditional aspects of the nonnaturals. For example, the seven good health habits, advocated by Belloc and Breslow [4] can be considered the modern equivalent to the "*res non-naturales*" (though with more emphasis on individual behavior).
[8] For a substantiation of this claim, see the recent study of J. Riley [37]. It is also the thesis of the well-known book of T. McKeown [30]. Already, in 1965, the Dutch physician Verdoorn proved by historical research that the nineteenth century decline of mortality in Amsterdam and the social dynamics of local culture were closely associated. Conditions of housing, nutrition, and hygiene have been of greater importance to the general improvement of health, since 1750, than advances in scientific medicine [44].

[9] Consider Bentham's famous dictum: "If we could suppose a new people, a generation of children: the legislator, finding no expectations formed which could oppose his views, might fashion them at his pleasure, as the sculptor fashions a block of marble" ([5], vol. 1, p. 323).
[10] This view is related to his Benthamite image of man. Man, according to Chadwick, is determined by his surroundings, but this environment is controllable and changeable. In the *Sanitary Report* it was concluded: "The course of the present inquiry shows how strongly circumstances that are governable govern the habits of the population, and in some instances appear almost to breed the species of the population" ([9], p. 164). This conclusion represents the 19th-century belief in Progress: man is perfectible through social engineering. It is mirrored in Bentham's preoccupation with eliminating contingency. Social phenomena are characterized by unpredictability and capriciousness and therefore uncontrollability. One way to eliminate their contingency is to make individual behaviour transparent. The greater its transparency, the more it is open to public control, and the more it will comply with social (i.e., utilitarian) norms (See [23]).

## BIBLIOGRAPHY


1. Ackerknecht, E. H.: 1948, 'Hygiene in France, 1815–1848', *Bulletin of the History of Medicine* 22, 117–155.
2. Bachelard, G.: 1938, *La formation de l'esprit scientifique. Contribution à une psychanalyse de la connaissance objective*, 11th ed. (1th ed., 1938), Librairie Philosophique J. Vrin, Paris.
3. Baumer, F. L.: 1977, *Modern European Thought: Continuity and Change in Ideas, 1600–1950*, MacMillan Publishing Co., New York.
4. Belloc, N. B. and Breslow, L.: 1972, 'Relationship of Physical Health Status and Health Practices', *Preventive Medicine* 1, 409–421.
5. Bentham, J.: 1843, *The Works of Jeremy Bentham*, J. Bowring (ed.), 11 vols., William Tait-Simpkin, Marshall & Co., Edinburgh-London.
6. Brandt, A.: 1987, *No Magic Bullet: A Social History of Venereal Disease in the United States since 1880* (2nd ed.), Oxford University Press, New York-Oxford.
7. Brockington, C. F.: 1965, *Public Health in the 19th century*, Livingstone, Edinburgh.
8. Burns, C. R.: 1976, 'The Nonnaturals: A Paradox in the Western Concept of Health', *The Journal of Medicine and Philosophy* 1, 202–211.
9. Chadwick, E.: 1965, *Report on the Sanitary Condition of the Labouring Population of Great-Britain*, M. W. Flinn (ed.) (original text, 1842), University Press, Edinburgh.
10. D'Aumerie, J. F.: 1833, *Herinneringen uit de cholera-epidemie te Scheveningen*, 's-Gravenhage.
11. De Bosch Kemper, J.: 1851, *Geschiedkundig onderzoek naar de armoede in ons vaderland*, Haarlem.
12. Donzelot, J.: 1977, *La Police des Familles*, Les Editions de Minuit, Paris; 1979, trans., *The Policing of Families*, Pantheon Books, New York.

13. Edelstein, L.: 1967, 'The Dietetics of Antiquity', *Ancient Medicine: Selected Papers of Ludwig Edelstein*, O. Temkin and C. L. Temkin (eds.), Johns Hopkins Press, Baltimore, Maryland, pp. 303–316.
14. Evans, R. J.: 1987, *Death in Hamburg: Society and Politics in the Cholera Years 1830–1910*, Clarendon Press, Oxford.
15. Fleck, L.: 1980, *Entstehung und Entwicklung einer wissenschaftlichen Tatsache: Einführung in die Lehre vom Denkstil und Denkkollektiv* (1935), Suhrkamp Verlag, Frankfurt am Main.
16. Fleck, L.: 1986, 'Scientific Observation and Perception in General' (1935), in R. S. Cohen and T. Schnelle (eds.), *Cognition and Fact: Materials on Ludwik Fleck*, D. Reidel Publishing Co., Dordrecht, Netherlands, pp. 59–78.
17. Fleck, L.: 1986, 'The Problem of Epistemology' (1936), in R. S. Cohen and T. Schnelle (eds.), *Cognition and Fact: Materials on Ludwik Fleck*, D. Reidel Publishing Co., Dordrecht, Netherlands, pp. 79–112.
18. Foucault, M.: 1973, *The Birth of the Clinic*, Pantheon Books, New York.
19. Foucault, M.: 1977, *Discipline and Punish. The Birth of the Prison*, Vintage Books, New York.
20. Fracastorius, H.: 1591, 'De contagionibus et contagiosis morbis et eorum curatione libristres' (1546), in *Opera Omnia*, Franciscus Fabrus, Lugduni.
21. Garrison, F. H.: 1910, 'Fracastorius, Athanasius Kircher and the Germ Theory of Diseases', *Science* 31, 500–502.
22. Have, H. A. M. J. ten: 1983, *Geneeskunde en Filosofie: De invloed van Jeremy Bentham op het medisch denken en handelen.* Uitgeversmaatschappij De Tijdstroom, Lochem.
23. Have, H. A. M. J. ten: 1986, *Jeremy Bentham: Een Quantumtheorie van de Ethiek*, Kok Agora, Kampen.
24. Have, H. A. M. J. ten: 1986, 'Gezondheidszorg en Samenleving: Over de idee van geneeskunde als sociale wetenschap', *Scripta Medico-Philosophica* **1,** 10–28.
25. Hippocrates: 1972, *Airs, Waters, Places*, in Hippocrates, Loeb Classical Library, trans. W. H. S. Jones, William Heinemann Ltd., Harvard University Press, Cambridge, Mass., vol. I, pp. 65–137.
26. Howard-Jones, N.: 1977, 'Fracastoro and Henle: A re-appraisal of their contribution to the concept of communicable diseases', *Medical History* 21, 61–68.
27. Howe, G. M.: 1976, *Man, Environment, and Disease in Britain: A Medical Geography of Britain through the Ages* (2nd ed.), Penguin Books ltd., Harmondsworth.
28. Jarcho, S.: 1970, 'Galen's Six Non-naturals. A Bibliographic Note and Translation', *Bulletin of the History of Medicine* 44, 372–377.
29. Kircher, A.: 1658, *Scrutinium physico-medicum contagiosae luis quae dicitur pestis*, Rome.
30. McKeown, T.: 1976, *The Role of Medicine: Dream, Mirage, or Nemesis?* The Nuffield Provincial Hospitals Trust, London.
31. Mill, J. S.: 1974, 'Utilitarianism' (1861), in M. Warnock (ed.), *Utilitarianism, On Liberty, Essay on Bentham. John Stuart Mill, together with selected writings of*

*Jeremy Bentham and John Austin* (11th ed.), Collins, Glasgow, pp. 251–321.
32. Morris, R. J.: 1976, *Cholera 1832: The Social Response to an Epidemic*, Croom Helm, London.
33. Niebyl, P. H.: 1971, 'The Non-naturals', *Bulletin of the History of Medicine* 45, 486–492.
34. Pelling, M.: 1978, *Cholera, Fever and English Medicine 1825–1865*, Oxford University Press, Oxford.
35. Pringle, J.: 1765, *Observations on the Diseases of the Army* (5th ed.), London.
36. Rather, L. J.: 1968, 'The 'Six Things Non-Natural'. A Note on the Origins and Fate of a Doctrine and a Phrase', *Clio Medica* 3, 337–347.
37. Riley, J. C.: 1987, *The Eighteenth century Campaign to Avoid Disease*, The MacMillan Press Ltd, London.
38. Rosen, G.: 1974, *From Medical Policy to Social Medicine: Essays on the History of Health Care*, Science History Publications, New York.
39. Rosenberg, C.: 1960, 'The Cause of Cholera: Aspects of Etiological Thought in 19th century America', *Bulletin of the History of Medicine*, 34, 331–354.
40. Shryock, R. H.: 1953, 'The Interplay of Social and Internal Factors in Modern Medicine. A historical Analysis', *Centaurus 3*, 107–125.
41. Shryock, R. H.: 1972, 'Germ Theories in Medicine prior to 1870: Further Comments on Continuity in Science', *Clio Medica* 7, 81–109.
42. Simon, J.: 1897, *English Sanitary Institutions, Reviewed in their Course of Development, and in some of their Political and Social Relations* (2nd ed.), Smith, Elder & Co., London.
43. Truffi, M.: 1947, 'The Story of Hieronymus Fracastor: Renaissance Physician and Originator of the Term "Syphilis"', *The Urologic and Cutaneous Review* 51, 515–533.
44. Verdoorn, J.: 1965, *Volksgezondheid en Sociale Ontwikkeling*, Het Spectrum, Utrecht-Antwerpen.

GERRIT K. KIMSMA

# FRAMES OF REFERENCE AND THE GROWTH OF MEDICAL KNOWLEDGE: L. FLECK AND M. FOUCAULT

## INTRODUCTION

The previous essay illuminates the historical development of medical scientific ideas. Medical science is shown to be *the result* of conflict and competition between various convictions, ideas, and ideological positions. Some of these struggles, at least temporarily, even seem to have been resolved by majority vote.

The reality of sickness and disease – in this particular case the epidemics of cholera – has served as a correction of misconceptions and as a guiding force toward more tenable, meaning: effective, ideas and theories. What has been made clear, too, is best illustrated by rephrasing George Rosen's conclusion: the development or growth of medical knowledge is highly dependent on non-intellectual factors and discoveries but even more so on political and economic interests [17].

In what follows I intend to offer an epistemological description in which the interrelationship between scientific discoveries, scientific ideas, and medical power are dealt with. I will do so by focusing on the ideas of the Polish physician, Ludwik Fleck (1896-1961) and the French philosopher Michel Foucault (1926-1984), viewing the work of both in relation to the work of Thomas Kuhn.

One certain conclusion concerning the growth of knowledge of infectious diseases, resulting in "the paradigmatic model" of infectious diseases for medical theory in general, can be put forward: it has been made abundantly clear that the growth of medical knowledge is not a matter of linear accumulation, continuously introducing novel corrections to misleading conceptions, leading to an ever tighter network of harmonious ideas.

One question arising from this conclusion is whether in the history of medical theories one can speak of a change of paradigm in the sense of Thomas Kuhn's view of the development of the physical sciences. That is, whether Kuhn's concept is applicable to medicine and leads to a better understanding of the development of medical science. On this subject there are highly interesting insights to share, especially since Kuhn has borrowed heavily from Fleck. Even the term 'paradigm' is more Fleckian than Kuh-



*H.A.M.J. ten Have et al. (eds.), The Growth of Medical Knowledge, 41–62.*

nian. I shall base my analysis on an early essay of Fleck as well as his now well-known *Genesis and Development of a Scientific Fact* [5].

The second question deals with the wider subject of how to understand the growth of medical knowledge if it is not the product of a strictly rational enterprise. This leads to an understanding of the growth of medical knowledge not as a logic of scientific discovery, but as a sociology of the historical situation of discoveries or changes in medical approaches to diseases like syphilis and cholera.

What we aim at is described in the title of Tsouyopoulos' article – we are searching for an adequate method for the history and theory of medicine [29]. Here again the ideas of Fleck are illuminating. Fleck's analysis of the history of the discovery of the Wassermann's reaction illustrates the need to introduce concepts like 'thought style' and 'thought coercion'. These concepts are needed to explain continuity in concepts where change was expected but where changes did not occur.[1]

Within this essay the ideas of Michel Foucault will be described, especially where they concern the understanding of specific frames of reference of medical thinking as a general *form* of thinking, as a discourse in society in general. Where thinking in its turn is part of a discourse, conceived as a unity of speaking, thinking and acting, and as an expression of a structuring force, it is designated by the Greek term 'episteme'.

Although it may seem to be the case at first, the present study is *not* historical, though it must be granted that the goal of understanding the growth of medical knowledge historically is in itself worthwhile. My aim is to try to clarify the growth of medical concepts and theories without a claim to historical completeness, but in order to search for underlying *ideas* that may account for the growth of medical knowledge.

The concepts of frames of reference and paradigms *of* or *in* medicine, however, are not only of historical or philosophical interest, but bear relevance for present-day medicine and its self-understanding. Henk ten Have's closing sentence – that medicine itself is a specific manifestation of history and culture [17] – signals that this specific manifestation, this medical enterprise, is in a state of flux and under serious critique from many non-medical social institutions such as law, economics and politics. Not only are the manifestations of the profession – e.g., status, income, and professional autonomy – in dispute, but even more so are medicine's concepts, goals, efficacy, and social role.

One of the major points of critique of the present-day medical enterprise, at least in the 'Old' and 'New' worlds, is the use of the prevailing medical model, i.e., the paradigmatic infection model, for solving social problems. Today, these problems are considered unmedical and beyond the scope and responsibility of the medical profession itself [32]. The same medical model that proved to be successful in conquering infectious diseases and was used to cure social diseases, is at the same time considered insufficient to deal with modern diseases, e.g., cancer, heart disease, and osteoporosis in the elderly [16].

The prevailing 'paradigm' is therefore not only under discussion regarding its application in unjustified areas of society, but also for its insufficiency in dealing with society's current diseases. Here it is interesting and illuminating to realize that the same concept that had been found usable for solving some of society's problems, is deemed insufficient to resolve current problems. This concerns the *moral value* of the medical model, that apparently was viewed as neutral, signifying that a social consensus on this strategy actually existed [36].

Henk ten Have's closing remark – "what is 'medical' or 'social' is not objectively demarcated" – becomes relevant since 'medical' and 'social' are terms that do not belong to the same category; rather, society determines what is 'medical'.

### *Kuhn's Notion of 'Paradigm' Revisited*

The Dutch physician and theoretical pathologist, H.S. Verbrugh, repeatedly stressed the insufficiency of contemporary medicine, because of its narrow conceptual base, both theoretically and in practice [30, 31, 32, 33]. In his work he applied Kuhn's notion of *paradigm* to medicine. 'Paradigm' refers both to a *system of thought* and to a *community of scholars* who adopt this narrow conceptual base without questioning the fundamental notions concerning the foundation of medical science and its true object. Applying Kuhn's notion, Verbrugh describes *three* paradigmatic periods: the Greek-Galenic, the Renaissance, and the era of modern medicine based on pathological anatomy, first indicated in Rudolf Virchow's *Cellular Pathology*.

He further distinguishes *two* contemporary periods: the humoral pathology, of the "closed body", and the pathology of the "open body" based on organic pathology. This echos the work of the well-known Dutch psychiatrist, J.H. van den Berg, who originally introduced these terms [2].

Verbrugh makes a plea for the restoration or at least integration of these two conceptual schemes. Humoral pathology has several advantages: it is holistic, expressed through the micro-macrocosmos terminology, and typically appears in therapeutic advice to patients. It allows for a different experience of corporality, overcoming the Cartesian dichotomy of body and mind, thus coming closer to a 'natural' experience. Humoral pathology also makes possible direct observation not inhibited by theoretical distinctions. And lastly, humoral pathology allows for a positive evaluation of the body's own healing power, making room for the personal responsibility of patients to listen to and follow their own natural healing processes and powers. So, in effect, a *revitalized* humoral pathology could signal a new system of reference for medicine, one that even encompasses organic pathology.

It is important to realize that the concept of paradigm seems to have received universal acceptance: its application to medicine seems beyond question. But this claim has not been thoroughly investigated. Recall that Kuhn developed his ideas in the context of the physical sciences. The application of the notion of paradigm to medicine rests on the presupposition of the critical identity between the physical sciences and medicine. In view of what has been stated, this remains a focal point in understanding not only the development of medicine but medicine as such. No one will challenge the obvious difference between the 'physical' and the 'social'; this is a basic distinction that requires little discussion. But this distinction is not so clear in medicine. Kuhn's ideas pertain to an adequate description of the tri-fold development of scientific knowledge: (1) the preparadigmatic, (2) one or more phases of a paradigm, and (3) one or more revolutionary phases [20, 21].

The question becomes one of determining whether Kuhn's ideas apply to medicine on the basis of a symmetry between the natural sciences and medicine. At this point the relevance of Fleck becomes important, notwithstanding the fact that the intellectual relationship between Kuhn and Fleck is somewhat ambivalent, though it should be noted that Kuhn might have saved Fleck from oblivion by mentioning his name in the foreword to his *Structure of Scientific Revolutions* ([20], p. VII).

Interestingly, the concept of *paradigm* originates with Fleck and made a detour from medical science to the natural sciences, and perhaps back to medicine again.

When describing his intellectual development Kuhn remarks that "... through it I have encountered Ludwik Fleck's almost unknown monograph, *Entstehung und Entwicklung einer wissenschaftlichen Tatsache* (Basel, 1935), an essay that anticipates many of my own ideas .... Fleck's work made me realize that those ideas might require to be set in the sociology of the scientific community ..." ([20], p. VII).

## LUDWIK FLECK (1896-1961)

### *Medicine* vs. *the Natural Sciences*

In one of Fleck's early essays he clarifies his remark that the differences between medicine and the natural sciences are more profound than the similarities [6]. These can be summarized as follows:

1. Medicine has no point of reference nor a unitary conceptuality. In opposition to ideas on the relationship between observer and object in the physical sciences, medical observation cannot be conceived as a fixed point. They are best conceived as a *circle*, because there is no single concept from which logically all phenomena of a disease can be described and categorized in one conceptual system. This means that the theory of medicine is always partially rational and irrational. In medical theory different and sometimes mutually excluding explanations are used. Sometimes causal explanations, sometimes teleological or pathogenetic explanations, even an account of the constitution of patients may be used to explain a course of events or the course of a disease. This appears to be accomplished in quite an arbitrary fashion. The reason for this lack of unity, at the same time and within the same framework, is due to the multitude of disease phenomena. There are no two patients with exactly the same disease or with exactly the same physiological or pathological characteristics. The multitude of clinical facts does not pose a problem for the inquiring mind if there is no real physical threat, but becomes problematic when there is a serious complaint or dysfunction.

2. The physician must employ the distinction between the *normal* and the *pathological.* Here a second major difference between the medical and physical sciences appears. Whereas the physical sciences establish laws governing the normal nature of things, medicine concentrates on deviations from the norm and works, in fact, with two concepts of normality [25].[2]

Medicine attempts to establish order out of chaos by working with "disease-pictures", based on a *constructed* theory of abnormality, *constructed* cause and effect relations, and on the simultaneous appearance of symptoms by constructing a syndrome. This formulation of disease-pictures calls for a high degree of abstraction, both from the individual patient, his or her life-history or from the relationship between organs and disease signs. Indeed, statistics are often employed to establish the 'ideal-type of a disease'.

3. In reality these disease-pictures are never quite congruent with the individual patient. Hence a third difference between medicine and the natural

sciences. In medicine it is possible to observe a phenomenon that is incompatible with the physician's existing theory without causing him to abandon that theory. Numerous are the para- and pseudo-diseases. Here the discrepancy between theory and practice is a fact of daily life, causing the creation of subclasses of abnormality and leaving the structure of classification and the taxonomy of disease intact. In effect, to establish a semblance of rationality, different concepts are declared to be related while they are of a decidedly different nature. For the same disease one may find causal and teleological explanations, external causality as well as internal, serving as compensation for the atypicity of the object of medicine. In Fleck's words: "This is what one encounters in the case of any medical problem: it becomes ever and ever necessary to alter the angle of vision, and to retreat from a consistent mental attitude" ([6], p. 43).

4. Medical concepts closely approach those of the physical sciences at the point of the *historia morbi*, the disease history. However, the concept of the history of a unique event is relevant to medicine but nowhere visible in the natural sciences, where exactness pertains to collections of signs and symptoms under observation. Also, the *historia morbi* reveals a double concept of time: first, time as a mode of change of normality; second, time as the pathological change of normal states-in-motion and related to each other as are acceleration and normal movement.

In short, application of the Kuhnian notion of paradigm in the natural sciences cannot simply be applied to medicine due to the marked and fundamental differences between these two human enterprises. Medicine has therefore no single paradigm or frame of reference from which it can adequately grasp all phenomena rationally and consistently ([6], p. 46).

### *The Genesis and Development of a Scientific Fact*

In *Genesis and Development of a Scientific Fact* [5] the question of understanding the conceptual development of medicine is reconsidered and approached as a social event, to be reconstructed as a sociological project in order to understand the final outcome.

The central event in Fleck's book on the history of syphilis is his discussion of the discovery of the Wassermann reaction in 1906. For Fleck this discovery was revolutionary because it caused a fundamental change in the physician's understanding of syphilis. The discovery also serves as an illustration of the *illogical*, *political*, and *irrational nature of scientific development*. This development is considered *illogical* because it is impossible to describe it as the result of logical premisses; it is called *political* because the research was instigated by the civil authorities; and it is *irrational* because the result cannot be described in terms of a successful succession of problems to be solved. On the contrary, its history reads like a series of pitfalls and is based on false presuppositions. Its history is not reproducible. Some lines of research that at first showed no sign of progress, later on yielded positive results. Fleck states that the discovery – or the creation – of the Wassermann reaction took place in a unique, historical process, that is neither experimentally reproducible nor logically reconstructable. He makes this statement as an insider, as one who has been there. His conclusion is that the only explanation why the Wassermann reaction was discovered at all was the drive and moral indignation created by syphilis research. It is this social climate that formed the collective effort, creating through continuous effort a *thought collective, das Denkkollektiv* as Fleck calls it, that resulted in the mutual and anonymous production of this particular reaction ([5], p.104). Surprisingly but characteristic of the power to maintain a self-image of the thought collective those involved with the research described the development as a *logical* process, carefully thought through and well-planned every step of the way. That is, the scientific community allows its members to forget the actual events, the failures, and to retain the ideal of a history of heroic research.

From Kuhn's position the Wassermann reaction can be seen as a revolutionary leap and has both social and epistemological consequences. Its significance is not reducible to the mere ability to discover who is infected.

Epistemologically, it is not merely new but revolutionary knowledge because it caused the prevailing concept of syphilis to change fundamentally. Not only does 'new' knowledge appear but 'old' knowledge becomes reshuffled. The Wassermann reaction forged a link between hitherto separate diseases by grouping syphilis' stages I, II and III (dementia paralytica) together. It also caused a narrowing down of this concept by separating it from diseases formerly thought to be associated with it, like phtisis, lupus, and rachitis, thus stimulating further independent research on these diseases. The Wassermann reaction enabled the cause of syphilis to be narrowed down to the bland spirochaete – the spirochaeta pallida. This narrowing down to a single cause created an openness regarding the concept of causality because the former absolute relation between the presence of the spirochaete and the disease was shattered as a result of the discovery of symptomless carriers.

The discovery of the Wassermann reaction caused the research to concentrate on identifying the real agent; as a result the spirochaeta pallida was discovered, but only *after* the reaction. Ironically, Fleck states that "... the discovery of the spirochaeta pallida is the result of the quiet labor of civil employees" ([5], p. 24). However hard to believe this was the case. Besides that, the Wassermann reaction meant the independence of a new medical field: the science of serology had become mature, not in the least because of its institutionalization and public funding.

Fleck's reflections on the conceptual consequences of the discovery of the Wassermann reaction reveal a sharp distinction between his and Kuhn's thinking. Whereas Kuhn notes a change in paradigm due to a revolutionary discovery or publication, Fleck illustrates convincingly the *continuity* among scientific concepts when seen in historical perspective in spite of the emergence of a revolutionary fact. The new knowledge does not signal a change in the frame of reference, but rather leads to a restatement of the "old" knowledge and ideas involved; this is clearly not a devaluation of old ideas, that made good sense in their historical context.

In his description of the history of syphilis Fleck distinguishes *four* separate and related ideas that are used to describe the phenomena involved in a disease:

(1) a mystical-ethical idea, relating syphilis to a particular stellar constellation and to the act of fornication – the "Lustseuche".
(2) an empirical-therapeutic notion, linking all venereal diseases that reacted positively to metals like mercury, antimone, or other metallic composites.
(3) a pathogenetic concept, based on the notion of "perverted blood", related directly to the broader and ancient concept of humoral pathology.
(4) the notion of a specific cause of syphilis – the etiology.

According to Fleck's description, the Wassermann reaction caused a change in the formulation of each of these concepts, but did not result in their disappearance. Syphilis is still related to the idea of sinful fornication, with a change in blood, and due to a specific cause. On closer analysis, the ideas of perverted blood and the unitary etiology were related to the moral indignation of the sinful character that prompted the political motivation to intervene, even though, as Fleck remarks ironically, the damage caused by tuberculosis was far greater.

This development is not to be considered a rational process of the growth of medical knowledge, though in effect it reflects an increasing rationality, predictability, and control; in essence it is a social proces. Scientific knowledge, and this is Fleck's original position, is essentially a social product, a social progeny, formed by moral, political, and other social factors [23]. This formation is executed through a thought collective, a group of scientists practicing a specific thought style [*Denkstil*], defined as the coherence of style of all – or many – concepts of an epoch and based on their mutual influence.

So opposed to a view of history as the growth of great ideas and the appearance of heroic personalities, Fleck stresses the *collective* nature of scientific knowledge, not demeaning the individual, but giving priority to the collectivity of scientists, to the thought style *over* individual thinking, and status to the collective over the individual scientist. But most of all there exists an indissolvable relation between the concept of science and its historical, psychological and political conditions: "... at least three quarters or maybe the total content of science is dependent upon or to be explained through the history, psychology, and sociology of scientific thinking" ([5], p. 24).

A thought style is not just a matter of rationality and change; it is a network of proto-ideas, scientific conceptions and values, has a dynamics of its own, one that reveals a resistance to change, a '*Beharrungstendenz*', as Fleck calls it. This is one explanation of the historical continuity that explicitly contradicts the idea of growth of knowledge as a process of fundamental rupture, since thinking reveals a certain "physiology of thought", that points toward a structural aspect of the human mind. This structural aspect escapes the agility of the human mind and can be described as the *economics of conceptuality*: new ideas are always formed within a certain conceptual force field of present concepts, and there tends to be resistance to massive change. The economics of concept change shows a preference for the theory that explains the most phenomena and with little need for different theories as is possible. At the same time, however, the higher its capacity to explain, the more resistant to change the theory will be. This is especially true for the concept of syphilis as carnal plague, as *Lustseuche*, because it inhibited further research for a long time by stressing the *moral* nature of syphilis. Once it was politically recognized, it prompted renewed research leading to the Wassermann reaction.

Fleck's conclusions from the history of the Wassermann reaction are many. He distinguishes in the history of ideas classical periods and periods of complications. In the classical period science grows by accumulations; the thought style functions as a coherent force, guiding individual observations and resisting change of existing theories. Fleck formulates a law such that what is "... allowed is only that which causes a minimum of arbitrariness of thought and a maximum of cohesion" ([5], p. 124).

That cohesion, however, is not just an affair of the scientist involved in the thought style, but is also dependent upon the general public. *Esoteric* and *exoteric* groups are related. Exoteric are those who are outside the scientific community but are somehow related to it, either as recipients or consumers; esoteric is the community of scientists. The cohesion of a given thought style or frame of reference is maintained on the basis of trust established between these two groups. Related to this are the 'open' or 'closed' character of a thought collective: the more secluded or closed its framework, the more frequent is the intracollective exchange of ideas that reinforces existing ideas and concepts. These ideas can now be applied to the present crisis in medicine. The less open to popular influence (or popular understanding, or trust) are the concepts of the esoteric group, the larger will be the complications for its dominant framework of ideas. According to Schäfer and Schnelle, Fleck sociologizes the theory of knowledge and historicizes the theory [28]. But there is more to it: Fleck is able to show, in close detail and with unmatched richness, the internal consistency and development of scientific ideas – at least in the field in which he was an insider – by underscoring the historical and sociological base of the growth of medical knowledge.

His ideas on the relation between popular journal, and handbook-science, are interesting as ways to close the gap between the eso- and exoteric groups within the thought collective, and also because of his keen insight regarding the change of terminology and the position of the individuals involved. There is more involved here than sociology or history; there is also a psychology of scientific ideas. Unique in Fleck's epistemology is the use of the terms *'active'* and *'passive'* in relation to knowledge. The understanding of the exact meaning of 'new knowledge' can be probed by its use. For example, the choice to unite all venereal diseases under the rubric *'Lustseuche'*, or carnal plague, is, in Fleck's terminology, an active linkage that cannot be explained either historically or psychologically. Both active and passive linking of specific contents of knowledge are expressions of a deeper phenomenon, the *Denkzwang*, the 'thought coercion' of the 'thought collective'. Compared to the active knowledge of *Lustseuche*, the knowledge resulting from the treatment of all these united diseases by ministrations of mercury or other metals (sometimes healing, sometimes making things worse) is called 'passive knowledge', because it is necessarily the result of the former link. One can say that the growth of medical knowledge implies the establishing of passive links based on some *actively* chosen connection. The advantage of this distinction, according to Fleck, is that it *overcomes the dualism between subject and object*, where the objectivity of knowledge frequently implies freedom from subjectivity. The idea of emotion-free thinking is unacceptable and its truth can never be established. Objectivity (i.e., the existence of formal systems of thought) includes human emotions and experience. The suggestion that certain concepts are free from emotions, e.g., the concept of causality, merely points toward a convention or unreflective consensus. Again, these phenomena of consensus are considered expressions of the *Denkzwang*. In general, each verbalization is a composite of active and passive knowledge; "No single sentence can be constructed from passive links; always an active part, or as it is called 'subjective', is present" ([5], p.68).

## MICHEL FOUCAULT (1926-1984)

Whereas Fleck describes science as a social product and illustrates his thesis through the Wassermann reaction and the essential tension between old and new frames of reference, Foucault describes the change in social consciousness that causes a change in the social network. In fact, where Fleck uses the terms '*Denkstil*' and '*Denkzwang*' in medicine to explain the tension between continuity and discontinuity, stressing the continuity of knowledge over a small period of time, Foucault describes the *discontinuity* of medical thinking as a direct result of a fundamental change in politics and social orientation.

Furthermore, Foucault offers an analysis of the relationship between what Fleck calls a 'thought collective' and 'thought coercion' by investigating the relationship of knowledge to power – or what he later calls the 'government' of individuals in society.

In *Birth of the Clinic* Foucault describes an epistemological "rupture" within medicine that occurred by 1800, the period of the French Revolution [11]. He distinguishes the medicine of the 'Classical Period' from that of 'Modern, Positive Science'. What on the surface of history appears as *continuous* cannot but be described as *discontinuous* at a deeper level, because the change in medical thinking is not only theoretical, but involves a complete turnabout affecting the relations between physicians and patients, between the state and individuals, and within the entire organization of medicine, both within hospitals and the medical profession within society.

Fleck had introduced the metaphor of the circle to describe the relation between observer/physician and observed object/patient/disease. In contrast,

Foucault introduces the notion of 'space' and delineates three different spaces. In the first place, there is the classification of diseases, thus the theory of medicine; secondly, he addresses the human body; thirdly, there is a social space – society as a whole. In his description of the epistemological rupture, the third space holds priority over the other two. This is important regarding the "crisis" in medicine and the question whether it is an epistemological or a political problem. In Foucault's words: "Tertiary is not intended to imply a derivative, less essential structure than the preceding ones; it brings into play a system of options that reveals the way in which a group, in order to protect itself, practices exclusions, establishes the forms of assistance, and reacts to poverty and to the fear of death .... In it a whole corpus of medical practices and institutions confronts the primary and secondary specializations with forms of a social space whose genesis, structure, and laws are of a different nature." Furthermore, " ...for this very reason, it is the point of origin for the most radical questionings. It so happened that it was on the basis of this tertiary spacialization, that the whole of medical experience was overturned and defined for its most concrete perceptions, new dimensions, and a new foundation" ([11], p. 16).

In general, the theory of medicine and the changes that occurred toward the end of the eighteenth century can be summarized as follows: for the seventeenth-century physician, diseases are entities in themselves, possessing their own laws and dynamics. The system of diseases reads like a flat two dimensional table, where each disease has a specific place. What is called a disease is entirely different from what it is called today: the distinction between signs and symptoms is rather arbitrary. There are families of diseases, genera and species, each branching out, based on a common denominator and a specific characteristic. A disease, for example, is a cramp, expressing itself differently in the various parts of the body. This system appears rational: it is a logically closed construct, where time, as in the history of a disease, is reduced or transformed into a step in the succession of logical steps of development of disease. First, there is a rough stage, stadium *cruditas*; then the increasing stage, stadium *incrementi*; then a *crisis* leading to a decreasing (stadium *decrementi*), or to death. The central element in this classification is the *resemblance* of a sign on the patient with the sign in the table, and the logical relationship is that of *analogy*. Between a slow progressive paralysis and a quick paralysis following a stroke there is essentially no difference. The human body is the field of expression of disease entities, the place where a disease becomes visible. Between diseases and the organism are regional and situational connections, but they concern only the sectors where a disease transports its specific dynamics. The organs "express" diseases that have an existence independent of the organs or the body. The organs are just the "incidental" carriers of a disease, because a disease can express itself in any part of the body. That is why a very exact anatomy is unnecessary and one need only know the general characteristic of the liver or intestines. The relationship between a disease and the body is ambivalent (at least for the physician), because it is a relationship of a pure disease to its deformation or disturbance through the body. So a patient causes a disturbance of a disease, through age, character, or disposition. The pathological fact is not considered anti-natural in relation to life, but rather the patient is anti-natural in relation to the disease itself. Medical activity is considered a deed of violence that interferes with the natural course of a disease, and any medical interference, either too early or too much, disguises the distribution and evolution of symptoms.

Both physician and patient are tolerated in view of the natural course of a disease, because they pose the matter of disturbance of the natural course. The position of patient is even more negative: to know a disease one necessarily has to abstract from the individual characterization of the patient, because at that level the individual is just a negative, albeit necessary, element. The fundamental epistemological act is therefore one of classification. In a symptom a disease is recognized; in a disease a species; in a certain species the pathological universe. This classification is justified because life and disease are not opposites but expressions of the same order: the rationality of life is identical to the rationality of that which threatens it. They are not related as nature and counter-nature, but exist in a mutual order in which they overlap and interact. In the disease, life is recognized because the law of life forms the foundation of disease: both are expressions of the one *divine creation.*

The change of this scheme of ideas toward the clinico-anatomical medicine of the nineteenth century is prompted by a critical re-evaluation of the position of hospitals in French society, and a change in the empirical approach of epidemics. It provokes also the first phase of socialization of medicine and its integration with the state. Influenced directly by the French Revolution, with its critical evaluation of hospitals as hatchingplaces for disease, the idea appears that medicine is an affair of the state, leading to the realization of the need for knowledge in order to be able to control disease. This in its turn leads to an investigation for health-endangering factors, be they climatological or personal, and thus a system of registration and a corps of registrars and inspectors. This empirical turn of medicine implies *a sharp departure from the medicine of classification.* The need for observations also means the development of systems of control of the medical establishment and a further integration of medicine and the state, in its turn resulting in debates over the quality of medical education, the position of hospitals, and the kind of medicine practiced within these institutions. These combined actions *cause a change in the position and value of medical knowledge in society*: the need to educate produced both a system of education and state interference in medicine, thus also shaping in the process the professionalization of the practitioners of medicine and elevating their social status.

What Foucault underscores is the absence of an adequate epistemological system, a frame of reference, in terms of which to execute the intentions of control of disease and the teaching of clinical medicine in hospitals. For positive medicine to emerge, a *mutation of the episteme* had to occur, one that involves a change in the relation between seeing and knowing, and the "vision" of the human triad: life, disease, and death. Episteme being the central concept in Foucault's "archeology" of the human sciences [8, 10] and disciplinary practices in dealing with the mentally insane [9] and the sick [11] and meaning the epistemological "grid" of relations that unites in a given period the (discursive) practices that make knowledge possible.

Strange as it may sound, the *medical body* now becomes the focal point of orientation. Instead of a negative image, the body becomes the field where the medical gaze orients itself in a new evaluation of symptoms and signs. The symptom is not the essence, but it refers to a hidden inside, the darkness of the human body.

A division is generated between the clinic and pathological anatomy, a division based upon the idea that death has no meaning for the clinic since it disturbs the signs of disease even more than in life. Death obscures symptoms and is the credo of the 'old' clinic. Secondly, the changes of disease must not be understood as logical steps within a temporal order, but as spatial changes *in* tissues and organs. The frequency of symptoms is not important for the 'new' clinic, but rather the establishment of a fixed point of reference in the body. In the case of tuberculosis that "point" is the destruction of lung tissue. So what belongs to the evident nature of clinical practice today, relating the exterior signs to the interior changes in function, involves a dramatic conceptual change. The chronology of symptoms is recognized as derivative from the topology of lesions. For modern medicine it is the body itself that is diseased; it is no longer a vehicle to express essences. It is essentially the stage of disease itself and the place where disease and life confront each other.

This change in medical theory involves a change in the role of death in medicine. From an event at the end of life, death becomes entangled *with life* where its signs are present in the living body in many forms and expressions. Death becomes a partner in a new trinity of life, disease, and death. Death even becomes the analytic point from which the relations of life in the living body become evident. The process of dying reveals the composition of body tissues by slowly decomposing the threads of life, and revealing a hierarchy of importance of the several organs in their ending sequence when they no longer function. Death becomes the mirror of life. Hence the need for autopsies in order to know the real disease: "open up a few corpses", as the expression of Bichat reads, is more than a slogan; it is the program for clinical medicine.

The fundamental characteristic of the relation between physician and patient in clinical anatomy, between the observing eye and the object of medicine, becomes the *invisible visibility*, in one-to-one relations, where each symptom corresponds with one word and one cause. Medical knowledge becomes knowledge of an interior that relates to the exterior as with the unveiling of a hidden object. Disease is visible because essentially the lesion needs to express itself. This involves a change in the evaluation of the division of signs and symptoms. The symptom becomes the medically relevant sign, and it is relevant only through the medical gaze. In the medicine of classification no sharp division existed between signs and symptoms, because any sign could become a symptom depending on its position in the table of pathology; but now a sharp differentiation is noted because a symptom can remain hidden and be discovered only after death. From a relation between *symptoms and disease* the essential relation becomes that of *symptoms and lesions*. This movement to the *interior of the body* leading to the invasion of formerly judged to be private parts is greatly enhanced by the use of instruments, as becomes apparent with the popularity of the stethoscope.

Between Kuhn's paradigm, Fleck's thought style, and Foucault's episteme are similarities and differences. A short description can aid both the understanding of each of these thinkers as well as make their relative positions more salient.

Contrary to Kuhn, Fleck stresses the irrational nature of scientific growth, and the continuity in conceptual changes, even though there appear no 'revolutionary' changes in a Kuhnian sense. Contrary to Fleck, Kuhn stresses the rational nature of existing paradigms, even though the change from one paradigm to another is not itself deemed a rational process. For Fleck, there are in a given thought style elements that are irrational in themselves and dependent upon society, be "society" conceived at large (the exoteric group) or smaller (the esoteric group of the scientists involved). This irrationality, for example, is found in the term 'thought coercion'. Compared to these notions, a similar characteristic can be described to clarify the relationship between Kuhn's paradigm and Foucault's episteme. One fundamental difference between Kuhn's paradigm and Foucault's episteme is the level at which each functions. Central for Foucault's concept is its "subliminal" or "subconscious" position, as opposed to the conscious status of a paradigm [24]. For Foucault, an episteme lies deeper and is significant not only for *one* science but for the central "grammar" governing several different human activities – life, labor, and culture. For the same reason, an episteme is "less changeable" than a paradigm. A paradigm, as a collection of rules, is changeable, tied as it is to a progressing professional practice. For Kuhn, a "new" paradigm is more valuable than the previous one because it is capable of solving "more puzzles" than the "old" one, alluding to a closer approximation to "objective knowledge" as representative of reality itself. Some theories are better than others. Foucault, however, would never admit to the possibility of objective knowledge, or the possibility of "a yardstick" to measure "the quality" of an episteme or the existence of "objective knowledge".

Another fundamental difference between Foucault, Fleck and Kuhn is the relationship between rational *knowledge* and *power* structures within society. Dreyfus and Rabinow point out the apparent similarity between Kuhn's concept of 'normal science' and Foucault's concept of science as part of a scheme of 'normalizing society', a power tactics of domination ([14], pp. 197ff). This similarity is, however, only on the surface, precisely because of the insidious nature Foucault sees hidden within the accumulation of scientific knowledge.

The Afterword on "The Subject and Power" [14] is a synopsis of Foucault's position. In it, he links the intent of his studies to the modes by which, in our culture, human beings are made *subjects*. Of that "subjection" both power and the presiding "episteme" form an essential part. The relation he intends to clarify is that of power and individualization. Although individualization is generally thought to be a positive movement in society, especially related to individual autonomy, at a deeper level (according to Foucault) society develops modes of power admitting forms of individualization while at the same time denying other forms. The surprising result is that the same movement to free people from certain forms of oppression results in other forms of *domination*. Here resistance is "... not so much for or against the "individual", but rather ... a struggle against the government of individualization" ([14], p. 212). This implies that common medical activities – diagnosing, prescribing therapies, establishing the freedom from labor, or the right to social benefits and advice on matters of hygiene or sexual behavior – constitute categorizations of individuals. And exactly this categorization imposes a truth on individuals which they must recognize and that others must recognize in them.

It is this process of "making subjects" that is a form of *power*, however hidden it may be. The term 'subject' is, "... tied to one's own identity by a conscience or self-knowledge more hidden than the other meaning: being subject to someone else by control or dependence" ([14], p. 213). In short, subjectivity can and must be viewed as the result of medical, discursive practice. In this respect Foucault calls attention to the symmetry between present-day medical power and the pastoral power of the eighteenth century that gives meaning to the often-heard contention that medicine has taken over the social space of religion ([14], p. 215).

Between religious, disciplinary tactics and the strategies of medicine is a link in which medicine, due to its basic reliance on a notion of normality, becomes the queen of science. Where the medical concept of normality suggests a neutrality that the social sciences rarely possess or not at all, its strategy nevertheless is adopted by practitioners of other professions, like psychologists and social workers. In general, Foucault maintains that the mere presence of standards of normalcy tend to legitimize activities to normalize, sometimes even making it compulsory or mandatory to comply, in order to become recognized as part of a group. So the differentiation between health and disease based upon a standard of normality has implications beyond its own field and becomes part of a more general social process of normalizing techniques in developing, for example, standards of hygiene and social behavior. Discipline and normality appear to be two sides of the same coin: compliance with the procedures of the one strategy implies obeisance to the other.

## CONCLUSION: FURTHER DEVELOPMENTS IN MEDICINE

Returning to our original question concerning the growth of medical knowledge and the adequacy of present-day medical theory, we can draw the following conclusions:

First, the longing for a change in the present medical paradigm appears to be romantic, both from the perspective of Foucault and from that of Fleck and Kuhn. From Foucault's perspective there is no way to change the episteme rationally or through management of health or disease, precisely because these rules and programs are expressions of a "hidden", subconsciously positioned episteme. Medical theory and epistemology are intimately bound to society at large. On the one hand, changes in society can mean or are capable of causing changes in medical thinking. On the other hand, the integration of medical theory, medical practice, and the ideology of the state, causes medicine to be both the vehicle of these aspirations and fears and also the executioner of new power relations. Thus medicine, as a discursive practice, with an exterior and an interior, a surface and a hidden character (in episteme and thought coercion) in its turn becomes the producer of a certain type of society, with certain types of power relations, and specific types of individualization. Thus medical theory and the institution of medicine as a whole are intimately connected with the realization of certain social values, again with a double face. The value of solidarity for example, as a direct descendent of the ideals of freedom, equality, and fraternity, at one level means a restatement of the Christian tradition of charity, but now for all and not just the happy few. While at another level it involves a hidden, but nevertheless quite present system of control and discipline, defining new power relations and inequalities, social control and alienation.

Given the intimate relationship between knowledge and power, Kuhn's notion of paradigm seems naïve and superficial. Fleck argues for the reality of social forces to shape science and medical practice, even to the extreme that a certain development (e.g., the history of the Wassermann reaction) cannot even be reconstructed as a rational enterprise that allows for the domination of irrational forces.

If the epistemology of medicine is socially changeable and ideology can cause an institution to be established or a profession to be recognized, the question of medicine's real point of reference becomes salient. Formulated in another way, one might conclude that medicine has no real point of reference, or at least that medicine's orientation toward health and disease is subject to other orientations or points of reference that may not be visible in their time but only through historical investigation.

On the other hand, the universal temptation of the Western medical enterprise and Western medical knowledge point toward a supercultural characteristic of medicine, standing above specific cultural values and leading to the conclusion that medicine is its own frame or reference in a fundamental sense and that it is auto-referential.

Fleck, who is far more optimistic about possible changes in thought collectives and thought styles than Foucault (probably because of the absence of the relation between power and knowledge in his work) views the relation of physicians and patients as being bound in circular discursive strategies. In his essay on "The Subject and Power" Foucault maintains that the way to understand power relations today is to analyze the "... forms of resistance against different forms of power and to use this resistance as a chemical catalyst, so as to bring to light power relations, locate their position, find out their point of application and the methods used" ([14], p. 210).

Resistance to medicine is both directed against its narrow view of disease as finite, physical entities and against the *domination* of the professional over the lay person. However, the possibilities for change effected by that resistance cannot simply be classified as a real change in the paradigmatic nature of medicine.

This can be illustrated from the patient's perspective. Arney and Bergen describe both an act of disappearance and of return of the 'experiencing person' in medicine [1]. 'Disappearance' means that for clinical medicine *subjective* experiences are of a second order of importance, whereas clinical abnormal functions and signs are of the first order. As the 'return' they mention the increasing clinical attention to emotions involved in sickness and the process of dying, as seen in the integration of psychology and social work within hospitals. The demand to be recognized as a complete human being, or "to die with dignity" may be interpreted as a rebellion of patients; yet "a new logic of truth speaking" emerges, being a transition within the structure of medical discourse that not only allows the patient to speak as an experiencing person, but *needs, demands* and *incites* him to speak. Thus the medical professional must come to manage emotions as well. This 'return' also signals a further extension of management, control, and discipline.

Where changes in medicine can be described as a *return* of the experiencing person, at the same time there are other developments that can be characterized as a further *disappearance* of the patient. This is typically said of developments in radiology and the field of prevention. In radiology, the development of the CAT-scan and the NMR-technique produce new fields of *visibility* in pathology, for which no adequate medical interpretation as yet exists, because the pictures cannot be recognized; there is an absence of a linear relation of the radiological and the clinical picture. Furthermore, abnormalities are discovered for which there not only are no names, but neither is there a complaint by the patient. The NMR presents an even further development in this respect: here the information is computerized, and can be manipulated visually even in the absence of the patient. Not only does the experience of the patient not matter, but neither is his *presence* necessary.

Preventive medicine presents the similar view. Here, the visibility expands while the experience of the patient diminishes, and the rationality of control is extended. Advice and guidance is freely supplied, based on a statistical prognosis of future, possible abnormalities, while no experience of disability or discomfort exists at all. So the rationality of control, the management of the living, is justified in the absence of a complaint, in effect repeating the medical concept of the body as the field of *potential* abnormalities. It is therefore not surprising that the advocates of prevention often present themselves as an *alternative* to clinical medicine. Regarding the disappearance of the person as an experiencing being no reform seems to be in sight, reinforcing the already small chance that there will be a change of paradigm in medicine.

*Westzaan,*
*The Netherlands*

## NOTES

[1] A study has been undertaken by G. Böhme to pursue the reason why medicine during the revolution of 1848 in Western-Europe did not become 'social medicine'. (See: "1848 und die Nicht-Entstehung der Sozialmedizin – über das Scheitern einer wissenschaftlichen Entwicklung und ihre politische Ursachen, *Kennis en Methode* 1979, 3, 119–141).

[2] With this "double concept of normality" is meant an essential medical fact: there are physiological *and* pathological deviations from the norm. Physiological deviations are by definition non-pathological and pathological are by definition non-physiological. This confers a "plasticity" on the theory and practice of medicine, enabling, in contrast to the physical sciences, the possibility that an observation that contradicts an existing theory does not end in a "crisis".

## BIBLIOGRAPHY

1. Arney, W.R. and Bergen, B.J.: 1987, *Medicine and the Management of the Living*, University of Chicago Press, Chicago.
2. Berg, J.H. van den: 1961, *Het Menselijk Lichaam*, Vol I and II, Callenbach, Nijkerk, Netherlands.
3. Berg, J.H. van den: 1963, *Leven in Meervoud*, Callenbach, Nijkerk, Netherlands.
4. Dreyfus, H.L. and Rabinow, P. (eds.): 1982, *Michel Foucault; Beyond Structuralism and Hermeneutics*, University of Chicago Press, Chicago.
5. Fleck, L.: 1979, *Entstehung und Entwicklung einer wissenschaftlichen Tatsache*

(1935), Suhrkamp Verlag, Frankfurt am Main; 1979, trans., *Genesis and Development of a Scientific Fact*, University of Chicago Press, Chicago.
6. Fleck, L.: 1986, 'Some Specific Features of the Medical Way of Thinking' (1927), in Cohen, R.S. and Schnelle, T. (eds.) 1986, *Cognition and Fact: Materials on Ludwik Fleck*, D. Reidel Publishing Co., Dordrecht, pp. 39–47.
7. Fleck, L.: 1935, 'Zur Frage der Grundlage medizinischen Erkenntnis', *Klinische Wochenschrift* 14, 1255–1259; 1981, trans., 'Ludwik Fleck's 'On the Question of the Foundation of Medical Knowledge', *Journal of Medicine and Philosophy* 6, 237–256.
8. Foucault, M.: 1966, *Les Mots et les Choses*, Gallimard, Paris; 1973, trans., *The Order of Things: An Archeology of the Human Sciences*, Vintage/Random House, New York.
9. Foucault, M.: 1961, *Folie et déraison; histoire de la folie à l'age classique*, Gallimard, Paris; 1973, trans., *Madness and Civilization; A History of Insanity in the Age of Reason*, Vintage/Random House, New York.
10. Foucault, M.: 1969, *L'archéologie du Savoir*, Gallimard, Paris; 1972, trans., *Archeology of Knowledge*, Harper and Row, New York.
11. Foucault, M.: 1963, *La Naissance de la Clinique*, Presses Universitaires de France, Paris; 1975, trans., *The Birth of the Clinic*, Vintage/Random House, New York.
12. Foucault, M.: 1976, *Mikrophysik der Macht*, Merve Verlag, Berlin.
13. Foucault, M.: 1981, 'Foucault over Macht', *Te Elfder Ure* 29, SUN, Nijmegen, 559–587.
14. Foucault, M.: 1982, 'The Subject and Power', in Dreyfus, H.L. and Rabinow, P. (eds.): 1982, *Michel Foucault: Beyond Structuralism and Hermeneutics*, University of Chicago Press, Chicago.
15. Foucault, M.: 1986, 'Waarheid als Macht', *Te Elfder Ure* 37/38, SUN, Nijmegen.
16. Greaves, D.: 1979, 'What is Medicine? Towards a Philosophical Approach', *Journal of Medical Ethics* 5, 29–32.
17. Have, H.A.M.J. ten: 1990, 'Knowledge and Practice in European Medicine: The Case of Infectious Diseases', in this volume, pp. 15–40.
18. Heelan, P.A.: 1986, 'Fleck's Contribution to Epistemology', in Cohen, R.S. and Schnelle, T., (eds.): 1986, *Cognition and Fact: Materials on Ludwik Fleck,* D. Reidel Publishing Co., Dordrecht, pp.287–309.
19. Illich, I.: 1976, *Limits to Medicine*, Marion Boyars, London.
20. Kuhn, T.: 1971, *The Structure of Scientific Revolutions*, University of Chicago Press, Chicago.
21. Kuhn, T.: 1977, *The Essential Tension: Selected Studies in Scientific Tradition and Change*. University of Chicago Press, Chicago.
22. Lakatos, I. and Musgrave, A.: 1970, *Criticism and the Growth of Knowledge*, Cambridge University Press, Cambridge.
23. McCullough, L.B.: 1981, 'Thought styles, Diagnosis and Concepts of Disease: Commentary on Ludwik Fleck', *Journal of Medicine and Philosophy* 6, 257–261.
24. Merquior, J.G.: 1985, *Foucault*, Fontana, London: 1988, trans., *De Filosofie van Michel Foucault*, Het Spectrum, Utrecht.

25. Oderwald, A.K.: 1985, *Geneeskunde in het teken van de semiologie*, Acco, Leuven.
26. Procee, H.: 1978, 'Thomas S. Kuhn over incommensurabiliteit', *Kennis en Methode* 2, 208–228.
27. Rolies, J. and Frijns, R.: 1987, 'Michel Foucault en de Geneeskunde', *Scripta Medico-Philosophica* 3, 35–39.
28. Schäfer, L. and Schnelle, T.: 1980, 'Ludwik Flecks Begründung der soziologischen Betrachtungsweise in der Wissenschaftstheorie'; Introduction to [5], pp.VII–XLIX.
29. Tsouyopoulos, N.: 1982, 'Auf der Suche nach einer adaequaten Methode für die Geschichte und Theorie der Medizin', *Medizinhistorisches Journal* 17, 20–36.
30. Verbrugh, H.S.: 1974, *Geneeskunde op dood spoor*, Lemniscaat, Rotterdam.
31. Verbrugh, H.S.: 1978, *Paradigma's en Begripsontwikkeling in de Geneeskunde*, De Toorts, Haarlem.
32. Verburgh, H.S.: 1983, *Nieuw Besef van Ziekte en Gezondheid*, De Toorts, Haarlem.
33. Verbrugh, H.S.: 1985, *Op de Huid van de Tijd*, De Toorts, Haarlem.
34. Vries, G. de: 1982, 'De Ontwikkeling van Wetenschappelijke Kennis', *Kennis en Methode* 6, 190–220.
35. Vries, G. de: 1981, 'De Besmettelijkheid van Intellectueel Contact', *Kennis en Methode*, 5, 156–164.
36. Zola, I.K.: 1975, 'In the Name of Health and Illness: on Some Socio-Political Consequences of Medical Influence', *Social Science and Medicine* 9, 83–87.

H. TRISTRAM ENGELHARDT, JR.

# MEDICAL KNOWLEDGE AND MEDICAL ACTION: COMPETING VISIONS

Much of the energy of contemporary epistemology has been focused on the interplay of facts, theories, and general views of reality. What is observed is recognized against a field or web of significance. In some sense, one must know for what one is looking before one can see it. Theoretical assumptions guide our gaze and frame our experience. All of this is true about the unapplied sciences of which medicine is *not* an instance. In the case of medicine, as with all applied sciences, values interact robustly with facts, theories, and visions of reality. In applied sciences some facts are more important than others, not because they help us to know more truly, but because they help us to treat more effectively, whether effectiveness is measured in monetary or non-monetary terms. As Henrik Wulff has argued, physicians are not so much concerned whether classifications are true or false, but whether they are useful [16]. Physicians tend to be pleased with an artificial, instrumental classification even if it does not provide a "true" picture of reality, as long as it maximizes the effectiveness of therapeutic interventions. In the applied sciences, epistemic goals are important, but non-epistemic goals are central. One frames understandings of the world within an applied science such as medicine not in order to know the world truly, but in order to control the world easily and cheaply. Under such circumstances true knowledge is in the service of effective action and the relationship between knowledge and power becomes explicit. It is important to note that communities of scientists, depending on whether their interests are primarily applied or unapplied, will vary in their views regarding what it means to know truly or to intervene effectively. This leads to a caveat presented by the work of Fleck [6] and Kuhn [12]. In talking about thought-collectives, thought-styles or paradigms, one must identify the particular community of scientists to which one wishes to make reference. Each community is defined by its own rules of evidence, inference, and negotiation. Scientists who are members of a particular community will in general know which facts are important and why. They will in general know when particular facts warrant particular conclusions or particular interventions. In the case of applied sciences, such as medicine, the rules of inference will include recipes for action, which will incorporate implicit, and at times explicit, value judgments



*H.A.M.J. ten Have et al. (eds.), The Growth of Medical Knowledge, 63–71.*

about proper trade-offs between costs and benefits. Also the character of proper trade-offs will often be negotiated without claiming that correct answers can be discovered. One might think here, for instance, of clinical classificatory schemes for cancers which form the basis not only for prognoses, but also for different levels of therapeutic interventions.[1]

As a result, caution is in order when talking about paradigms and thought-styles. One must identify particular communities of investigators or practitioners, and then map out in detail the rules of evidence, inference, and negotiation each community shares. As I have suggested elsewhere, a community sharing a thought-style or paradigm can be identified by its capacity to resolve a controversy by appeal to rational arguments (i.e., as when a dispute turns on epistemic issues in an unapplied science), or by its capacity to resolve a controversy by appeals to procedures for negotiating a common understanding (i.e., as when a dispute turns on non-epistemic issues as in an applied science) [4, 5]. It is very unlikely that as many paradigms separate as unite physicians. For example, paradigms of coronary artery disease, and exemplars of its cure, will bind together surgeons on the one hand, and internists on the other, and exclude either side from easily agreeing on the merits of competing surgical versus non-surgical approaches to treatment.

The history of medicine, as the foregoing papers have shown, is very important for understanding the character of contemporary medicine. History reminds us that medicine has been medicalizing complaints for centuries and that the worries of Ivan Illich [9] and others [10] do not address new difficulties, if there is a difficulty. Medicine has always been a powerful resource for controlling complaints. However, it is only recently that medicine has become an explicit governmental and societal endeavor in the proportion it is today, with the percentage of gross national product devoted to health care now ranging from five to eleven percent in most industrialized countries.

The history of medicine also reminds us that there are many paradigms of infectious diseases which have changed and altered over time, influenced in part by the development of paradigms elsewhere in science. One might think here of the importance of our contemporary understanding of the genetic code for our ability to account for the different infectious capacities of different varieties of RNA and DNA viruses. Our appreciation of infectious diseases has also been placed within the concerns of occupational medicine, epidemiology, and public health, which in different ways and through different institutions have addressed the social aspects of infectious diseases.

One might consider as a recent example the attention to the role played by bisexual life-styles in the transmission of acquired immune deficiency disease (AIDS).

In drawing out the implications of the rich papers by Henk ten Have [8] and Gerrit Kimsma [11], one must carefully chart the fine grain of paradigms in the institutions that sustain the various undertakings placed under the rubric of medicine. When one examines actual controversies in medicine, one may find that particular individuals involved in a controversy may be embedded in more than one thought community. They will not share the same rules of evidence, inference, and negotiation, and therefore the capacity to resolve a controversy with all the individuals involved in the controversy.

Thus, psychiatrists, as members of the American Psychiatric Association, may agree about how to describe a disorder but disagree about its cause. They may agree on how to use the Diagnostic and Statistical Manual [2], but not on the significance of that to which it is applied. Here one must again underscore the cardinal difference between the applied and unapplied sciences. Communities of physicians will be separated not only by epistemic interests, but by non-epistemic interests. The non-epistemic interests will include not only professional prestige (a matter that may separate teams of unapplied scientists as well), but economic advantage which is secured through particular therapeutic approaches (again even here there will be analogies; individuals engaged in an unapplied science who compete for limited research funds may have an economical and professional stake in discrediting the claims of competing groups of investigators).

Though there is much in Fleck [6], Kuhn [12], and Foucault [7] that helps us to account better for the interplay of facts, values, theories, and social structures, there is much obscurity as well. Kuhn does not help us by advancing a single clear account of what he means by paradigm. In the 1969 postscript to *The Structure of Scientific Revolutions* he offers a sociological definition, as well as one he claims to be the philosophically deeper of the two, which turns on exemplar solutions to concrete puzzles. "On the one hand, it [paradigm] stands for the entire constellation of beliefs, values, techniques, and so on shared by the members of a given community. On the other, it denotes one sort of element in that constellation, the concrete puzzle-solutions which, employed as models or examples, can replace explicit rules as a basis for the solution of the remaining puzzles of normal science" ([12], p. 175). The sociological account of paradigm, it should be noted, is cardinal. "A paradigm is what the members of a scientific community share, *and*, conversely, a scientific community consists of men who share a paradigm"

([12], p. 176). Kuhn appreciates that a paradigm belongs to a community of individuals who see problems in a particular way and share assumptions that allow for their solution. They have a common vision of reality and a common understanding about how that vision can be enlarged and tested.

Despite Kuhn's attempts to give a clear account of his notion of paradigm, it goes aground on the many and diverse ways in which he uses the term. In fact, Margaret Masterman enumerates twenty-one different usages of paradigm in Kuhn's *The Structure of Scientific Revolutions* [13]. She makes the important contribution of clustering the various senses under three rubrics: (1) metaphysical paradigms, which say something about the geography or furniture of reality; (2) sociological paradigms, which involve appeals directly or indirectly to the social and political structures that hold communities together under a common vision; (3) construct paradigms, which identify particular texts, tools, or guiding solutions to problems. She appreciates that communities of scientists are bound together by shared assumptions concerning the nature of reality and by common understandings of what constitutes instructively successful experiments, as well as by being members of the same professional associations and reading the same journals.

These analyses of the nature of thought-styles, thought collectives, and paradigms have led to a substantial reassessment of the line between discovering and creating reality. The traditional distinction between the context of justification and of discovery has been blurred as well. One tends to see that which can be justified. Theoretical expectations function both implicitly and explicitly to sort message from noise in the attempt to move from poorly structured to well structured problems. In this way, the context of justification influences the context of discovery. Moreover, what is derived within the context of discovery influences views of successful justification, for the fine grain of justification itself changes through time under the pressure of historical, cultural forces and previous successes at discovery.

All of this is to underscore that it is not possible to separate facts and visions of reality, though these two dimensions of knowledge are surely distinguishable. How one sees facts is determined by a wide range of both epistemic and non-epistemic concerns. Assumptions that true knowledge is marked by maximum simplicity are guided by quasi-aesthetic values regarding the nature of reality and of true knowledge. Non-epistemic values concerning the relative costs of over- versus under-treating clinical problems influence how reality is portrayed in recipes for intervention. On the other hand, facts do shape, challenge, and even disconfirm theories. An appreciation of the historical, social, and contextual character of knowledge claims does not doom one to a limitless relativity with regard to knowledge claims in general, or claims regarding competing visions of man's relationship to nature, in particular. However, one is compelled to analyze carefully the different roles played by considerations that tend to be more factual (i.e., more oriented around epistemic values) and those more oriented around epistemic and non-epistemic values.

I will use Thomas Sowell's term "vision" and recast it to compass an extended sense of what holds communities together in a particular understanding of reality [15]. This also requires an expansion of what Fleck meant by thought-styles and thought-collectives. My goal is to provide a term that compasses not only what holds communities of scientists together, but also the communities of discourse framed around particular moral, aesthetic, religious, and political commitments. Individuals who share a vision share sufficient rules of evidence, inference, and negotiation to resolve controversies regarding the central challenges and problems they see within their domain of common interest. As I have noted, individuals may, and usually if not always do, belong to more than one community of discourse. As a result, conflicts and confusions can arise, not only for members of different groups when they meet, but also for individuals who are members of more than one community (and who, as a result, may be subject to significant cognitive and affective dissonance). In such circumstances often no answer can be discovered, and at best one can be created by some arbitrary process.

Visions of the world provide five clusters of leitmotifs for understanding and acting on the world. First, they provide axiological guidance. They tell what to value and disvalue and in what order. In the area of the unapplied sciences, axiological guidance focuses on epistemic values and includes views regarding what it means to know truly (e.g., what counts as mathematically elegant portrayals of reality). In the area of the applied sciences, such as medicine, one is provided with hierarchies of values and harms that determine answers to such questions as what likelihood of mortality, morbidity, and financial risks or costs should be shouldered to avoid what likelihood of harms. Here one may find value commitments that answer how much economic burden must be shouldered in order to avoid what probability of significant harm to patients. Axiological guidance will also disclose the goals or values to be realized in medicine. Second, visions provide ontological guidance. They tell us about the ultimate structure of things. Here are included basic presuppositions about nature, reality, and causality. One finds guidance as to whether one should expect the amount of energy and matter to remain constant in the universe or to come into existence *ex nihilo* without cause. It also includes views of what will count as causes of disease. The third component is sociological. It identifies what marks the boundaries of particular communities of vision; it shows how they are sustained by implicit instructions regarding who are strangers and who are members, and the ways in which one should react to each. Under this component, one finds everything from professional associations and journals to binding codes of medical etiquette. Insofar as knowing and acting are communal undertakings, one requires a social fabric with social rules for joint endeavors within a vision. Fourth, there are examples of what it means to know things correctly. Here are included tools and successful experiments. Finally, if the vision is not that of a group of unapplied scientists, one will also find various warrants for intervention. Communities of applied scientists have implicit or explicit recipes indicating when it is prudent or imprudent to intervene in particular ways to alter the world. In medicine this is captured under the general notions of usual and customary standards of care or by formally articulated indications for treatment. This may also include portrayals of ideal relationships between physicians and patients, which function as inspiring, mythic exemplars for action.

Knowledge grows as capacities for the intersubjective resolution of disputes regarding the character of the world grow. The growth of medical knowledge can be seen as equivalent to growth in the capacity reliably over time discovering or creating answers to controversies regarding how most effectively to treat patients. The resolution of controversies has been closely tied to more explicit attention being given to the role of observer bias. In this regard one should note that classifiers such as Sauvages [14] and Cullen [2] emphasized caution in giving accounts regarding the relationship of clinical reality and pathoanatomical states. "Progress" was tied to seeing how clinical and pathoanatomical (as well as etiological) claims can be interrelated in ways open to empirical, intersubjective, critical analysis and falsification.

But the development of criteria for objectivity (as at least intersubjectivity) had not displaced the experiencing subject. Nor is it quite correct to say that modern medicine has displaced the experiencing patient. Nuclear magnetic resonance and other highly technological ways of examining the anatomical and physiological functioning of patients disclose diseases only insofar as the findings can be associated with the complaints or actual experiences of patients. Findings are pathological, rather than merely physiological variations, only insofar as they are correlated with problems that patients actually live through, actually experience.

Dissatisfaction with medicine is usually associated with controversies regarding what it means to *intervene* effectively rather than what it means to *know* truly. Patients and society may judge that medical institutions are more concerned with their own prestige, power, economic advantage or views of patients' best interests, than with how patients regard their complaints, concerns, and best interests. The problem then is not usually with rival accounts of the pathoanatomical or etiological underpinnings of patients' complaints and problems. Rather, the problem is how properly to respect the wishes and goals of patients within health care institutions that have their own particular hierarchies of values and concerns with regard to cost efficiency. Though at times alienation from modern medicine is grounded in patients' endorsing views of science or true medical knowledge different from their care givers, most alienation derives from a failure of medicine to take the wishes of patients seriously. Disputes focus primarily on what goals should be pursued through health care and how. This claim is substantiated by the growing interest in bioethics which has focused on patients' rights, on limiting paternalistic interventions by physicians, and on making explicit the values that physicians and health care institutions may unknowingly bring to the treatment of patients. The differences between the visions of medicine endorsed by patients versus those by physicians lie primarily in the axiological, sociological, and recipe-for-action related dimensions of those visions. Put another way, bioethics has focused on what distinguishes medicine as an applied science from the basic medical sciences as unapplied sciences.

The point is that all knowing is not the same. Claims made by applied and unapplied scientists differ in important ways. They differ in terms of the communities in which the claims are made and in terms of the rules of evidence, inference, and negotiation they endorse – all of which changes over and through time. Medicine is an important focus for the philosophy of science and technology because it so robustly illustrates the interplay of facts, theories, values, and power in an applied science where both epistemic and non-epistemic values are salient. For example, one does not simply want to know why a patient is dying, one wants to know in order to keep the patient from dying, make the dying easier, or at least in accord with the wishes of the patient. Generally, one seeks power to intervene on behalf of a rich matrix of non-epistemic concerns. It is for this reason that philosophical analysis in bioethics has been so closely tied to concerns with public policy: communities of patients or possible patients wish to participate in negotiating the recipes or indications for therapeutic interventions. Since physicians must act or treat, even when knowledge is imperfect, much must be negotiated rather than discovered. Physicians and patients must agree about how to act under conditions of uncertainty. The concern of patients is more often that they be a part of the community of negotiation regarding the application of medical knowledge (a social- and value-oriented interest), than that they have a part in establishing medical knowledge as such.

The instrumental and goal-directed character of medical knowledge means that there is not only an interaction between an explanans and explanandum, but also an interaction between a manipulans and manipulandum. In the case of medicine, that which is to be manipulated, the manipulandum, is we ourselves. Because medicine as a consequence touches on the whole range of human values and concerns, the occasion for controversy is obvious. Different communities of patients or potential patients with different hierarchies of values can have quite different understandings of the goals of medicine and how health care is appropriately to be given. The controversies are also often between different communities of practitioners separated not by, or not primarily by, different views about how to know truly, but by different views regarding what is involved in *acting* effectively and properly. A careful examination of these rich interactions will help us not just to understand medicine better, but applied sciences and technologies in general.

*Baylor College of Medicine,*
*Houston, Texas, U.S.A.*

## NOTE

[1] Clinical classifications that are used in staging cancer are negotiated, voted on, adopted, and promulgated with the general understanding that they are *instrumental* rather than *natural* classifications. They are recognized to be artificial construals of reality that serve the purposes of the clinician and the investigator.

## BIBLIOGRAPHY

1. American Joint Committee on Cancer: 1983, *Manual for Staging of Cancer*, 2d ed., Lippincott, Philadelphia.
2. American Psychiatric Association: 1980, *Diagnostic and Statistical Manual of Mental Disorders*, American Psychiatric Association, Washington, D.C.
3. Cullen, W.: 1769, *Synopsis nosologiae methodicae*, William Creech, Edinburgh.
4. Engelhardt, H.T., Jr., and Caplan, A.: 1985, *Scientific Controversies: A Study in the Resolution and Closure of Disputes Concerning Science and Technology*, Cambridge University Press, Cambridge, Mass.
5. Engelhardt, H.T., Jr.: 1986, *The Foundations of Bioethics*, Oxford University Press, New York.

6. Fleck, L.: 1935, *Entstehung und Entwicklung einer wissenschaftlichen Tatsache. Einführung in die Lehre vom Denkstil und Denkkollektiv*, Benno Schwabe, Basel; 1979, *Genesis and Development of a Scientific Fact*, (ed.) T.J. Trenn and R.K. Merton, trans. F. Bradley and T.J. Trenn, University of Chicago Press, Chicago.
7. Foucault, M.: 1963, *Naissance de la Clinique*, Presses Universitaires de France, Paris; 1973, *The Birth of the Clinic: An Archaeology of Medical Perception*, trans. A.M. Sheridan Smith, Random House, New York.
8. Have, H. ten: 1990, 'Knowledge and Practice in European Medicine: The Case of Infectious Diseases', in this volume, pp. 15–40.
9. Illich, I.: 1976, *Medical Nemesis*, Pantheon Books, New York.
10. Kass, L.: 1975, 'Regarding the End of Medicine and the Pursuit of Health', *Public Interest* 40 (Summer), 11–24.
11. Kimsma, G. K.: 1990, 'Frames of Reference and the Growth of Medical Knowledge: L. Fleck and M. Foucault', in this volume, pp. 41–62.
12. Kuhn, T.: 1970, *The Structure of Scientific Revolutions*, 2nd ed., University of Chicago Press, Chicago, Illinois.
13. Masterman, M.: 1970, 'The Nature of a Paradigm', in *Criticism and the Growth of Knowledge*, (ed.) I. Lakatos and A. Musgrave, Cambridge University Press, London, pp. 59–89.
14. Sauvages de la Croix, F. B.: 1763, *Nosologia methodica sistens morborum classes juxta Sydenhami mentem et botanicorum ordinem*, 5 vols., Fratrum de Tournes, Amsterdam.
15. Sowell, T.: 1987, *A Conflict of Visions*, William Morrow, New York.
16. Wulff, H.: 1981, *Rational Diagnosis and Treatment*, 2d ed., Blackwell Scientific Publications, London.

# SECTION II

# PHILOSOPHY OF SCIENCE AND THE GROWTH OF MEDICAL KNOWLEDGE

HENRIK R. WULFF

# FUNCTION AND VALUE OF MEDICAL KNOWLEDGE IN MODERN DISEASES

## INTRODUCTION

Imagine a medical scientist who is engaged in studying the very complex pathogenetic mechanisms of a certain disease process. He may well find that these studies are extremely satisfying from a purely academic point of view, and it is quite possible that it is this intellectual satisfaction which motivates him. However, as a medical scientist he will realize that it is the ultimate aim of all medical activities to promote health and to eliminate illness, and he will hope that one day his studies may serve that end.

We should look at medical philosophy in much the same way. Academically, it may be very interesting to analyze medical problems, but philosophy of medicine is also a medical discipline, and as such it is subservient to the same ultimate aim as medical science. It may be intellectually stimulating to analyze medical thinking, but those medically interested philosophers and philosophically minded members of the medical profession who engage themselves in such studies should see it as their purpose to improve the practice of medicine.

Therefore, the topic of this essay is a very important one. It invites a discussion of those fundamental problems of modern medicine which are the object of serious concern among many members of the medical profession.

As an internist, I am myself painfully aware of the limitations of contemporary medicine. We see patients with acute myocardial infarction, and although we can treat some of the complications effectively, we do not come to grips with the underlying disease process and we cannot effectively prevent recurrences. We see patients with cerebrovascular accidents whose future quality of life is bound to be very poor, and it is often difficult to decide how far we should go in terms of investigations and treatment. We see patients with cancers which we can treat but not cure, and we see an increasing number of alcoholics where our knowledge of medical science is of little use when we try to motivate them to stop drinking.

A few decades ago most people inside and outside the medical profession agreed that the development of modern medicine was a success story. They had seen the introduction of insulin, vitamin $B_{12}$, antibiotics, oral diuretics, new surgical techniques etc., and they took it for granted that progress would



*H.A.M.J. ten Have et al. (eds.), The Growth of Medical Knowledge, 75–86.*

continue. So it does in some areas, but on the whole it is as if medical science has lost some of its impetus. Are we facing problems which cannot be solved within the traditional framework of thinking? Thomas Kuhn has taught us that, during periods of paradigmatic unrest, scientists begin to wonder about the very basis of their thinking [3]. The present interest in the philosophy of medicine may be seen as a symptom of such unrest in the field of medicine.

## CLARIFICATION OF TERMS

Let us look once again at the words in the title of this essay. The expression *'modern diseases'* will simply be taken to mean those diseases which today dominate clinical practice in Western society – and some examples have been given already.

The term *'medical knowledge'* will also be used fairly loosely, comprising all that which the contemporary doctor must know in order to tackle the health problems which he or she encounters. It is the main purpose of this essay to try to analyze the foundation of clinical decisions.

*'Function'* is always function for a purpose, and, as mentioned already, it is the ultimate purpose of all medical activities to further health and to eliminate disease. This statement is perhaps somewhat trivial, but, as I shall explain later, it does hold the key to some of the problems that we face.

Then we must distinguish between ethical and non-ethical *'value'*, where non-ethical value is almost synonymous with effectiveness in relation to a given purpose. Here the word will be used mostly in its non-ethical sense, but even then it has ethical overtones, as the ultimate goal – the promotion of health or quality of life – must be considered "good in itself".

Finally, the words *'science'* and *'scientific'* will be used in their narrow sense, i.e., pertaining to natural science.

## BIOLOGICAL KNOWLEDGE

The first type of knowledge to be discussed is biological knowledge and by that is meant all the scientific knowledge which we possess as regards the structure and function of the human organism in health and disease. Traditionally, biological knowledge is considered by far the most important requisite of clinical decisionmaking: it tells the clinician which investigations to do in order to reveal the cause of the patient's illness, the cause often being the diagnosis. Then, when he has made the diagnosis, i.e., identified the disease from which the patient is suffering, he uses his biological knowledge to select that treatment, if any, which will eliminate the disease. According to this traditional view, disease is regarded as abnormal biological structure and function.

This framework of thinking was first established during the first half of the 19th century. Around 1800, French pathologists correlated autopsy findings to those symptoms and signs they had recorded when the patient was alive, and they felt that they actually *saw* the causes of their patients' complaints on the autopsy table. From their point of view diseases were anatomical lesions. Later that century, other medical scientists took an interest in abnormal function and initiated the development of physiology and, at a later stage, biochemistry. It was as if they had acquired new glasses which made them see *functional* disturbances where their predecessors had seen anatomical abnormalities, and they established new functionally-defined disease entities, such as hypertension, myxoedema (defined as decreased function of the thyroid gland) and diabetes mellitus (defined as disturbed carbohydrate metabolism). Other scientists wore neither "anatomical" nor "physiological" spectacles, but focused on microorganisms (bacteria) as causes of disease. To them it was self-evident that it was not the anatomical lesions (the tubercles), but Koch's bacillus which constituted the real cause of tuberculosis.

This development led to the present disease classification which is a mixture of these schools of thought. It is of paramount importance to clinical thinking as it serves to organize all clinical knowledge and experience.

One might have wished that it had been possible for the medical profession to classify their patients in a more systematic and logical way, but on the whole the end result has been quite successful. The diseases which we diagnose are often defined by the identification of a single causal factor in a complex causal network, but if this causal factor is a non-redundant component of the effective causal complex [4, 10], its elimination may well have a curative effect. The development of scientific medicine, which is closely linked to the development of the disease classification, has led to effective treatment of numerous diseases, and every day doctors save the lives of patients who almost certainly would have died a century ago. Why then cannot we expect that future acquisition of more *biological* knowledge will teach us to cure those "modern" diseases which dominate today?

First of all, the problems which we face today to a very large extent simply represent the reverse of the medal, – they are the inevitable effects of the success story. It has proved possible to cure previously lethal diseases which affected people in the younger age groups and, consequently, we must expect to see more patients with the diseases of old age. Further, it has proved possible to treat intercurrent infections and acute exacerbations in old people with chronic diseases, and, consequently, we must expect to see more old people in the advanced stages of those chronic diseases. I recently talked to an elderly nurse who in her younger days worked at a home for old people. She said that most of them were quite well, played cards, and visited their relatives, until they "suddenly" died. I frequently visit a nursing home in our neighbourhood and now the situation is very different: the majority of the patients are very old, helpless, and suffering from senile dementia.

What we see is the effect of antibiotics, oral diuretics, and other treatments, and of course we should not just deplore the development. We must not forget all those old people who now live satisfying lives in their own homes, when some decades ago they would have died.

Nor must we draw the conclusion from this picture that the present paradigm of medical thought is running dry. We still see important breakthroughs and more are to be expected. Ten years ago, the first effective treatment of peptic ulcer disease saw light, and some malignant diseases such as Hodgkin's disease and certain leukemias have recently sometimes proved curable. There is no reason to believe that we shall not learn to cure more malignant diseases, and there is nothing to suggest that it will not prove possible to find effective remedies against the most menacing of modern diseases, AIDS.

We may also see progress in the fields of, for instance, coronary surgery and transplantation surgery, but that will not change the picture radically. On the contrary, in future we shall have a better chance of surviving an acute myocardial infarction, but that also means that we shall run a greater risk of ending up at a nursing home in a demented state.

We are not likely to develop effective cures for the degenerative diseases of old age, e.g., treatments which may eliminate the arteriosclerotic vascular changes, and we must therefore turn our attention to the possibility of prevention.

Unfortunately, progress in prevention has in most areas been very much slower than progress in therapy. We have learned to prevent a number of infectious diseases, but otherwise our ability to prevent disease is still very limited. There is one good reason for that. The number of scientists who concern themselves with the environmental determinants of disease is extremely small compared with the number of those who study disease mechanisms [8]. I shall give two examples from gastroenterology: It is well established that gastric ulcer disease in the 1880s was a common disease among young women ([9], p. 62), whereas duodenal ulcer disease was very rare. Today gastric ulcer disease is frequent in middle-aged and elderly people of both sexes, and duodenal ulcer disease is common, especially in men. Such drastic changes in the panorama of peptic ulcer disease prove the importance of environmental factors of one kind or another. Nevertheless, the number of scientific papers dealing with this problem is minute compared to the number of those dealing with gastric acid production and other aspects of gastric and duodenal function.

Similarly, the incidence of Crohn's disease in Northern Europe has increased significantly during the last few decades, for which reason environmental factors must play an important role, but once again the number of studies dealing with possible etiological factors is minute compared with the number of studies dealing with the disease mechanism. We cannot expect nature to provide the answers when we do not ask the questions.

In this connection we must also remember the subtle point that the disease classification is constructed in such a way that it serves primarily a *therapeutic* purpose. If we look once again at the structure of the disease classification, it will be realized that it is based primarily on pathogenetic factors, i.e., on the identification of the "mechanical fault" in the complex human "machine". We have established such disease entities as iron-deficiency anaemia, myocardial infarction, and hyperthyroidism, and all these diagnoses are therapeutically valuable as they tell us what went wrong in the body and, perhaps, how we may remedy the "mechanical" fault, but they do not tell us how to prevent the fault. A group of patients with the same disease harbor, by definition, the same "mechanical" fault, but it is quite possible that the etiology, i.e., the *environmental* and *genetic* factors which initiated the disease process in these patients, presented great variation. In that case attempts to trace the causal network backwards from the "mechanical" fault to the etiological factors may be a wild-goose chase in the sense that the end result is bound to be the identification of a large number of so-called 'risk factors'. A disease can only be prevented if those people who suffer from the disease have one necessary etiological causal factor in common, and that does not seem to be true of most degenerative diseases.

The implication of these considerations is not that the "modern" diseases, including arteriosclerosis and malignant diseases, cannot be prevented, but rather that we may be attacking the problem in the wrong way. Rather than reasoning from effect to cause, we ought to reason from cause to effect. We know, for instance, that cigarette smoking may be one of the causative factors of a variety of different diseases, such as lung cancer, chronic bronchitis, bladder cancer, myocardial infarction, intermittent claudication, and duodenal ulcers, and it is quite possible that studies of other etiological factors will reveal an equally varied picture. Rather than trying to find *the* etiology of arteriosclerosis, we should perhaps study in much greater detail the effects of the things we ingest and inhale, as well as our life styles and living conditions, e.g., shift work, unemployment, and career pressure.

## CLINICAL EXPERIENCE

According to the traditional view, it is taken for granted that clinicians use their theoretical, biological knowledge to choose the best treatment, in much the same way as it is taken for granted that airplane engineers use their theoretical knowledge of aerodymanics when they design the shape of a new airplane. This view is partly right and partly wrong.

Theoretical knowledge is important, but so is the experience which is gained by testing airplanes in wind tunnels and by using the medical treatment in practice. When we discuss medical knowledge, we must also take into account the knowledge which is obtained by clinical experience.

This distinction between theoretical knowledge and clinical experience is, of course, closely related to the traditional distinction between basic science and applied science or technology. The physicist who studies the laws of aerodynamics is a basic scientist who generates new knowledge, whereas the engineer doing research in wind tunnels is a technologist, who is concerned with the solution of a practical problem.

Perhaps the distinction is somewhat less clearcut in medicine than in other fields, as it may be argued that by definition there is no such thing as pure medical science, medicine as such being defined by its aim, but there can be no doubt that the clinical research worker, who studies the effect of different treatments at the bedside, is a technologist to the same extent as the aircraft engineer. He does not worry too much about the mechanism of action, if only the treatment is safe and effective, in the same way as the engineer does not worry about our incomplete knowledge of aerodynamics, if only the airplane fulfills the requirements.

Clinical experience may be *uncontrolled*, which means that it was obtained in the daily routine, or it may be *controlled*, which means that it was obtained doing meticulously planned clinical surveys and trials.

Previously, doctors introducing new treatments relied solely on their biological knowledge and their uncontrolled experience. A treatment was thought to be effective, if it seemed rational and if it seemed effective in the daily routine. The history of medicine, however, shows that this mode of thinking is usually insufficient and often dangerous. In the middle of the last century, doctors treated severely ill cholera patients with bloodletting as it seemed effective in the daily routine and rational according to the theories of the time, and this disastrous treatment was used until the underlying (Hippocratic) theory was abandoned decades later. In the 1950s, the medical profession introduced anticoagulant treatment of acute myocardial infarction on a similar basis, and its ineffectiveness was not appreciated until the 1960s when the results of properly controlled trials were published.

It must be admitted that the effects of some treatments, such as streptomycin treatment of tuberculous meningitis, raw liver in pernicious anaemia, and insulin treatment of diabetic coma, was so pronounced that clinical trials were unnecessary, but these examples constitute the exception rather than the rule. Most new treatments which are introduced today are only marginally better than the current ones, and demonstration of their superiority requires randomized, preferably double-blind, trials [9].

Nowadays many members of the medical profession appreciate the importance of controlled clinical studies, but less critical colleagues still introduce new therapeutic methods of doubtful worth, and the benefits of many of those diagnostic technologies which we use have never been demonstrated.

Nevertheless, even controlled clinical experience cannot be separated from biological theory. Firstly, there is a constant interaction between the scientist doing laboratory studies and the clinical research worker who carries out studies at the bedside. The former generates those hypotheses as regards, for instance, the effects of new treatments which must be tested at the bedside, and the latter makes observations which give the scientists food for thought.

Secondly, the results of clinical trials are rarely clearcut. They usually provide numerical results that are compatible with both the null hypothesis (that the effect of the new treatment is equal to that of the old one) and the alternative hypothesis (that the effects of the two treatments differ). Therefore, the clinician who has to choose between the two treatments must take into account not only the results of the trial, but also his prior beliefs. These beliefs to a large extent reflect his theoretical knowledge, and, consequently, the interpretation of the results of clinical research is never theory-free. Research workers sometimes believe that they avoid these problems by doing statistical significance tests, but that, of course, is not true. The introduction of biostatistical methods has greatly improved the quality of medical research, but the p-values which constitute the results of the significance tests cannot not be interpreted except in the light of the prior beliefs of the research worker (or the reader of the medical journal) ([10], pp. 99–102).

This is not the place to discuss clinical research in any detail, but it must be stressed that in the case of the "modern" diseases there is a great need for an intensified clinical research effort. The number of clinical trials being reported in the medical journals is on the increase, which is a good thing, but unfortunately the collective research effort is heavily biased by commercial interests. There are numerous studies of the effect of a variety of new beta-blockers in myocardial infarction, but much fewer studies of old established treatments, as, for instance, the effect of digitalis in elderly persons with cardiac incompetence.

## HERMENEUTIC KNOWLEDGE

To this point clinical medicine has been discussed as if it was only a branch of natural science, and that is unacceptable. According to the extreme *scientific view* mankind is seen only as a biological phenomenon, disease being no more than abnormal biological function; clinicians see it as their goal to restore this function to normal. Thanks to their scientific knowledge, they are the ones who know which treatment is best, and their attitude toward the patients is that of extreme paternalism.

This picture of clinical medicine is, of course, a caricature, which is not and never was true, but, personally, I would not mind too much if my doctor treated me in that way in case I suffered from meningococcal meningitis or pernicious anaemia. My only concern would be that the doctor had enough scientific knowledge to make the right diagnosis and institute the right treatment.

But, probably, no clinician ever held the extreme scientific view. Medical practitioners have at all times realized that medicine is both an art and a science – that one must distinguish between scientific and humanistic medicine –, and thus it might be more correct to say that clinicians often adopt the scientific view *with humanistic constraints*. They adopt the scientific view of disease as abnormal biological function, but they also show compassion and treat their patients as fellow human beings. They feel that they know which treatment is best, but they also appreciate that they must respect their patients' autonomy and secure their informed consent.

Now, however, the gradual change in the disease spectrum is forcing the medical profession to adopt the much more radical position that some of the premises of the scientific view are false, especially the assumption that disease is no more than abnormal biological function and that the "goodness" of different treatments can be assessed objectively. I shall approach this problem from both a theoretical and a practical perspective.

Theoretically, it has, at least in my opinion, proved impossible to justify the definition of disease as abnormal biological function. Those who have tried to defend the biological disease concept, e.g., Boorse, Scadding and Ross [1, 6, 7] seem to run into serious difficulties when they try to define what is meant by normal biological function ([10], pp. 46–59).

Function is always *function for a purpose*, and the difficulty is seen most clearly if we look at the hierarchy of biological functions: The mitochondrion may be said to function normally, if it makes its proper contribution to the normal function of the cell, of which it is a part; the cell may be said to function normally, if it makes its proper contribution to the normal function of the organ, to which it belongs; the organ may be said to function normally if it makes its proper contribution to the function of the whole organism, etc. But what does it mean that the whole organism functions normally? In biological terms the purpose of life is the survival of the individual and the reproduction of the species, but from a medical point of view that conclusion is clearly unsatisfactory. The normality of the function of a human being must be viewed in relation to the life experience, hopes, values, wishes and feelings of that particular individual.

This discussion is rather theoretical when we are dealing with diseases such as acute appendicitis or pernicious anaemia, but it is of paramount importance when we are dealing with the modern diseases. Let us return to the problems faced by the contemporary internist on his daily wardround: Is it good or bad to treat an intercurrent infection in an old, aphatic, and hemiplegic person? Is it good to prolong the life of a cancer patient if the treatment has unpleasant side-effects? Decisions of this kind require a different kind of knowledge – *knowledge based on empathic understanding of that particular patient*, knowledge derived from dialogues with the patient and the patient's relatives.

In other less dramatic circumstances the clinician may have a choice between different treatments (e.g., different treatments of hyperthyroidism and peptic ulcer disease), and also in these cases the selection of the "best" treatment requires the active participation of the patient. Medical ethicists often discuss respect of individual autonomy *versus* clinical paternalism, and they usually regard this dilemma as a purely ethical one. It is often overlooked that *paternalism is the logical consequence of the biological concept of disease*, and that the rejection of that concept necessitates the participation of the patient in the decision process.

Clinicians also see patients who suffer from alcoholism and other forms of addiction, and in such cases biological knowledge is of limited importance. Knowledge of the effects of alcohol may help the clinician to make the diagnosis in the first place and to predict the outcome, if the patient continues to drink, but the only effective treatment is *motivation* to stop drinking. This is a trivial example, but it is worth noting that the word 'motivation' has little meaning from the point of view of the natural scientist. It belongs to the realm of humanistic medicine, according to which man is not only a biological, but also a self-reflecting being, who can choose in accordance with his own values.

As yet, medical research workers concern themselves almost exclusively with biological research and quantitative clinical studies, but there is a growing realization of the need for so-called qualitative research. There is no reason to believe that research methods from a variety of humanistic disciplines (history, linguistics, studies of literature, and music) cannot be applied to medical problems, so that medical research, just like medical practice, may reflect that medicine is both an art and a science.

Similarly, the medical philosopher should see it as one of his aims to bridge the gap between the Anglo-Saxon tradition of empiricist thinking and the hermeneutic schools of Continental philosophy.

## KNOWLEDGE OF ETHICS

A distinction has been made between scientific and humanistic medicine, and I have mentioned what may be called the hermeneutic component of medicine as a humanistic discipline, but that does not exhaust the discussion. Our scientific knowledge tells us what is wrong with the patient from a biological point of view and, to some extent, what will be the result of different treatments, but it does not necessarily tell us how we *ought* to act in a particular situation.

It is sometimes very difficult to decide whether or not it is justified to prolong the life of a patient, if it is certain that the future quality of life will be very poor. A problem like this one has a hermeneutic dimension, as the doctor, as far as possible, must take into account the patient's own wishes, but it also represents an ethical dilemma. The doctor must consider what is meant by his duty to save human lives. Does life in this context simply mean biological life, or does it mean that kind of life which permits the individual to express himself as a human being?

The doctor may decide, as it was once said that there is "... as much moral onus on us to determine that life-saving treatment is required in a particular instance as there is to determine that it is not" ([2], p.126).

Many doctors today avoid this dilemma by always advising active treatment and in some cases they even admit openly that they would have hoped that the patient had died, if he or she had been a close relative. This is just one example which shows that doctors today must be able to analyze complex ethical problems. They must consider the consequences of their actions for the patient (the patient-orientated utilitarian view), the general consequences of their actions (the universal utilitarian view), and they must take into consideration the patient's rights and their own duties (the deontological view) [10].

It might facilitate clinical decisionmaking, if we paid greater attention to the *empirical aspects of medical ethics*. Every society is characterised by an unwritten ethical code which helps to govern our activities with one another, and it would be useful to record which kinds of medical behavior is valued and expected by, for instance, Danish, Dutch, and American hospital patients. It is quite possible that such empirical studies would reveal pronounced differences. It is now generally accepted that medical ethics is an important *medical* discipline, but it is not always appreciated that it is not an international one, as, for instance, physiology or orthopaedics. Medical ethics is at least to some extent culturally specific, as it deals with the application of general ethical norms of a particular society to medical problems.

## KNOWLEDGE RELATED TO HEALTH ECONOMY

Clinical decisions require more than considerations for the particular patient in a particular situation. We must realize that resources in any health care system are limited and that we must always consider the *opportunity cost* of our actions, i.e., the opportunities which we lose in one part of the health care system by spending the money in another. In order to make such assessments the clinician must know the cost of the diagnostic and therapeutic methods which he or she uses.

New biological knowledge and new technological developments lead to the introduction of expensive diagnostic investigations and therapeutic methods, and we shall only be able to cope with the health problems of modern society with its focus on chronic incurable diseases if we distribute our limited resources as fairly as possible. The theory of John Rawls may be applicable to this problem [5].

In this way, the change in the disease spectrum also dictates that doctors possess some knowledge of health economics, and it is promising that contemporary health economists seem well aware of the ethical aspects of the new discipline.

In summary, the health problems which we face today force us to realize that clinical decisionmaking cannot be contained within the framework of scientific thinking. Scientific medicine, based on biological knowledge, is only a tool, albeit it a very important one, which serves an end that transcends the biomedical sciences.

*Herlev University Hospital,*
*Denmark*

## BIBLIOGRAPHY

1. Boorse, C.: 1977, 'Health as a Theoretical Concept', *Philosophy of Science* 44, 542–573.
2. Cohen, C.B.: 1983, '"Quality of Life" and the Analogy with the Nazis', *Journal of Medicine and Philosophy* 8, 113–135.
3. Kuhn, T.S.: 1972, *The Structure of Scientific Revolutions*, 2nd ed., University of Chicago Press, Chicago.
4. Mackie, J.L.: 1973, *The Cement of the Universe. A Study of Causation*, Oxford University Press, Oxford.
5. Rawls, J.: 1971, *A Theory of Justice*, Oxford University Press, Oxford.
6. Ross, A.: 1979, 'Sygdomsbegrebet', *Bibliotek for læger* 171, 111–129.
7. Scadding, J.C.: 1967, 'Diagnosis: the Clinician and the Computer', *Lancet ii*, 877–882.
8. Wulff, H.R.: 1980, 'Teaching Value of International Congresses', *Revista española de las enfermedades del aparato digestiva* 17 (suppl. IV), 91–93.
9. Wulff, H.R.: 1981, *Rational Diagnosis and Treatment*, 2nd ed., Blackwell Scientific Publications, Oxford.
10. Wulff, H.R., Pedersen, S.A., and Rosenberg, R.: 1986, *Philosophy of Medicine*, Blackwell Scientific Publications, Oxford.

PAUL J. THUNG

# THE GROWTH OF MEDICAL KNOWLEDGE: AN EPISTEMOLOGICAL EXPLORATION

## INTRODUCTION

This paper has been inspired by two contrasting views on modern medicine. One is the conviction, popular since at least some 100 years, that medicine is becoming progressively more effective. Drawing on a growing store of knowledge and technology, modern science will ultimately enable man to live out his natural life-time with a minimum of ill-health or disease. The other view is more recent and less optimistic. Since the late 1960s it has been observed that modern medicine often impedes effective health care, especially under conditions of poverty. It draws away means and manpower needed for the prevention and treatment of wholesale health-deficiencies of the general population, and spends them on sophisticated diagnosis and repair of the afflictions of the urban upper class [3, 5].

The contrast between these views is usually dissolved by distinguishing between the science of medicine on the one hand and the practice of health care on the other. Medical science is claimed to be a powerful instrument which, if applied correctly and in sufficient measure, could indeed solve most health problems. The practice of health care, on the other hand, is plagued by social and economic problems which frustrate the provision of appropriate medical care to the general population. Any reconciliation of the two views mentioned above implicitly assumes that medicine is competent, at least in principle, to solve health problems as presented by society. This assumption, however, is philosophically gratuitous and raises a number of questions: What justification have we for claiming the effectiveness of medical knowledge and technology? What type of problems does medical science solve and to what extent are these problems similar to the ones encountered in the practice of health care?

Addressing these questions, I will start from the fact that, since the second half of the 19th century, medical knowledge is mostly regarded as natural science applied to matters of health and disease. The introduction, since the 1950s, of psychology and sociology as added ingredients, although acknowledged by many general practitioners and psychiatrists, is practically ignored by most clinicians. Sections II and III therefore discuss natural scientific knowledge and its growing effectiveness. In section IV it will be



*H.A.M.J. ten Have et al. (eds.), The Growth of Medical Knowledge, 87–101.*

concluded that this growth is progressive as well as potentially unlimited.

Next it will be argued in section V that this conclusion, when applied to *medical* knowledge, holds true only as long as we accept the restricted definition of medicine as applied natural science. This definition, however, should be rejected since it restricts the range of medical competence to such problems as are framed in natural scientific terms. Section VI suggests that in our days an expanded, more realistic definition of medicine is emerging. The nature of this redefinition of medicine is sketched, as well as its implications for the future development of medical knowledge and the criteria to determine its growth and progress (section VII).

## EXPLAINING SCIENTIFIC PROGRESS: FROM REALISM TO AGNOSTICISM

Although the roots of natural science can be traced to antiquity, it is helpful to take as our starting point the "Scientific Revolution" between the late 15th and 17th centuries. That period saw the birth of experimental science, or in T. Kuhn's [7] more discriminating terminology, of the Baconian sciences. And it is this development which is usually regarded as the origin of the dynamic character of modern science, of its growth potential, and its progressiveness.

Crucial for this development was the increasing availability to the experimenters of physical and chemical apparatus. This development is often ascribed to the social and economical emancipation of technicians and craftsmen. Their traditions and skills were linked up with the views and arguments of philosophers in what Bernal calls "the marriage of the craftsman to the scholar." This marriage was to be fertile indeed: "the combination of the two approaches took some time to work out, and spread rather gradually at first through the different parts of knowledge and action. But once the constituents had been brought together there was no stopping the combination – it was an explosive one" ([1], pp. 261–262).

The intriguing question of course is: Why did this new approach engender such an explosive development? What explains its success? This question is often answered in terms of a naïve scientific realism which defines truth as the correspondence between scientific theories and nature, while positing nature as existing independently of human conceptualization ([11], p. 89). This position is implicit in the writings of scientists recording their satisfaction with what science has achieved and their expectations of further successes [10]. It entails a vision of cumulative growth and progress of the natural sciences in Western culture since the Renaissance. As criteria for progress, terms like '*explanatory power*' and '*predictive precision*' are invoked without further discussion, as well as the general notion of the '*mastery of nature*'.

*Scientific realism* is explicitly defended by Karl Popper in his evolutionary theory of scientific knowledge. Although he deems it neither demonstrable nor refutable, he accepts realism as essential to common sense as well as to science and philosophy, both of which he describes as "enlightened common sense" [13]. His justification for this position boils down to a biological argument:

> "As children we learn to decode the chaotic messages which meet us from our environment. We learn to sift them, to ignore the majority of them, and to single out those which are of biological importance to use either at once, or in a future for which we are being prepared by a process of maturation." Furthermore: "We are, I conjecture, innately disposed to refer the messages to a coherent and partly regular and ordered system: to 'reality'. In other words, our subjective knowledge of reality consists of maturing innate dispositions" ([13], p. 63).

Popper subsequently sketches an "Evolutionary Epistemology": truth is a biological fact, it means that our beliefs are adapted to our environment. "Starting from scientific realism it is fairly clear that if our actions and reactions were badly adjusted to our environment, we should not survive" ([13], p. 69).

Survival, in other words, demonstrates that our "innate dispositions" are adapted to our environment and this would seem to validate scientific realism. It should be noted, however, that the very notion of "adaptation to the environment" itself already implies realism as a theory of knowledge! For Popper, the circularity of this argument is unproblematic. It is impossible to validate realism; it can neither be demonstrated nor tested, but it is a plausible theory and it essentially appeals to common sense. Indeed, it is implicit in all scientific endeavour, and even our language, being largely descriptive, presupposes an objective reality. It is, in brief, the only sensible hypothesis and a conjecture to which no sensible alternative has ever been offered ([13], pp. 40–42).

The realist's definition and explanation of scientific progress is very simple and commonsensical, indeed. True knowledge delivers practical results because it reflects nature in an instrumental way. Its success lies in its power to manipulate reality, and it is clear that the experimental method, with its steering and feedback relations between theory and reality, is especially conducive to manipulative success. This also explains the cumulativeness of scientific progress: each technological gain enables us to further refine and extend our grip on nature.

Kuhn offers a more sophisticated version of scientific realism, which purports no metaphysical position. He denies that "truth about nature" is an adequate criterion for scientific progress, because we cannot recognize it. Yet Kuhn describes the growth of science as a process wherein theory is matched to nature: "With the passage of time, scientific theories taken as a group are obviously more and more articulated. In the process, they are matched to nature at an increasing number of points and with increasing precision" ([7], p. 289). This observation, however, should not be read as a naïve realism. As pointed out earlier by Kuhn [6], his position differs from that of most scientists and laymen because his belief in scientific progress carries no metaphysical implications. "A scientific theory is usually felt to be better than its predecessors not only in the sense that it is a better instrument for discovery and solving puzzles but also because it is somehow a better representation of what nature is really like. One often hears that successive theories grow ever closer to, or approximate more and more closely to, the truth. Apparently generalizations like that refer not to the puzzle-solutions and the concrete predictions derived from a theory but rather to its ontology, to the match, that is, between the entities with which the theory populates nature and what is 'really there'" ([6], p. 206). Kuhn emphatically rejects this latter implication. Successive scientific theories may show progress as instruments for puzzle-solving, but this does not reflect on their ontologies. It should be noted that this position dislocates Kuhn's earlier argument that science, throughout its *revolutionary* history, shows *evolutionary* progress ([6], ch. XIII). Of this problem Kuhn is well aware. The history of science is marked by revolutions during which the scientific community changes its paradigmatic tradition of problem-analysis, research, and explanation. The term 'paradigmatic' here refers to ([6], p. 10) "paradigm" as introduced in reference to those examples of scientific practice, including law, theory, application and instrumentation together, which provide models for successive traditions of scientific research. Kuhn's view on the historical development of science involves periods of so-called normal science, during which a common paradigm dominates which also implies that a common model of reality is adhered to. Scientific revolutions, however, marked by a shift of paradigms, engender a discontinuity in the ontological claims of scientific knowledge during successive historical periods. But if scientific knowledge has no ontological continuity, there is no anchorage for criteria of progress to bridge the successive scientific revolutions. Kuhn in fact concedes that he has not really answered the question why the evolutionary process in scientific development should work. "Why should consensus endure across one paradigm change after another? And why should paradigm change invariably produce an instrument more perfect in any sense that those known before?" He concludes that the ultimate question: "What must the world be like in order that man may know it?" has not been and need not be answered, because "any conception of nature compatible with the growth of science by proof is compatible with the evolutionary view of science developed here" ([6], p. 173).

## THE MATHEMATICAL DESIGN OF NATURE

Explanations of scientific progress quoted so far, left us somewhere between a pragmatic realism and an ontological agnosticism. For this ambivalent position, however, a solution was offered long ago in continental philosophy. Kuhn's "ultimate question" had in fact been tackled by Heidegger in 1935, but in reverse: What must human knowledge be like in order that it may relate to reality? I here refer to his "Die Frage nach dem Ding", tentatively translated as "The Quest for Reality" [4]. The fundamental prerequisite for experimental science, says Heidegger, is the mathematical design of reality. This expression does not indicate the actual calculus or geometry as applied in science, but the general principle or condition of all practical mathematics. Heidegger ([4], pp. 53 ff.) refers to the original sense of the Greek 'mathemata': what is learned and taught. It indicates things in as far as we take cognizance of them. Mathematics in this general sense is a mental construct; it is the design, the form and format, of man's experimental questions to nature and, simultaneously, of nature's empirical answers.

In modern science, natural objects are determined by this design. This is illustrated ([4], pp. 59 ff.) by confronting Newton's (and before him Galilei's) concept of space and movement with the Aristotelian view of the movements of physical bodies. Newton conceived of a homogeneous space in which things move as they are affected by forces, while in the earlier Greek world each body moved by virtue of and in accordance with its specific nature and place. There, the movements of celestial bodies were inherently different from movements of bodies on earth. Every "thing" was thus primarily determined by its nature, its somehow and somewhere. In its behavior it showed us this specific nature rather than a generalizable pattern. It is the Newtonian mathematical design which renders reality homogeneous and enables us to view it as ruled by universal laws. This modern view designs "things as things" ("Die Dingheit der Dinge"), as is evident from

Newton's first law which posits a body uninfluenced by impinging forces (*"corpus quod a viribus impressis non cogitur"*), a mental construct with no correlate in reality.

Discussing the metaphysical implications of the mathematical design of nature, Heidegger ([4], pp. 76 ff.) refers to Descartes, in whose "Regulae ad directionem ingenii" ("Rules for Guiding our Understanding") he recognizes an early description of the essence of modern science. In Descartes, modern thought broke free from antiquity, although the mathematical trend had already become apparent a century earlier. Descartes' "cogito" divested nature of its subjectivity: things no longer spoke for themselves (or for God, in the Christian variant of Aristotelianism). Once the "I" had been posited as the only and fundamental subject, all nature became its object and was laid open to the mathematical structure of human thought.

Heidegger pursues the metaphysical question from Descartes to Kant's philosophy ([4], pp. 92 ff.). For Kant, nature is the sum and content of all possible experience ("der Inbegriff des möglichen Erfahrbaren") and experience here refers to true knowledge ("gegründete und begründbare Erkenntnis"). This experience is founded on and structured by the basic principles of pure understanding ("Gründsätze des reinen Verstandes", in which "Verstand" indicates the judgment-function of "Vernunft" or reason). These principles themselves rest on the cooperation or unity of human thought and senses ("Denken und Anschauung"). This unity underlies all knowledge of the phenomenal world and Heidegger now sees the following circle as essential for Kant's epistemology: *"Die Grundsätze des reinen Verstandes sind durch dasjenige möglich was sie selbst ermöglichen sollen, die Erfahrung"* ([4], p. 174). Freely translated and related to what was said before: our knowledge of nature is shaped by principles of understanding which themselves are based on our experience of nature. And this leads to Heidegger's final conclusion, that man meets reality while he is himself the cause and condition of this meeting. The quest for reality is the quest for man, whose existence gives shape to a structure in which both reality and man himself are mutual determinants.

Now it should be remembered that this structure is the *mathesis universalis* of Descartes which returns in Kant's dictum that natural science is truly scientific only in as far as it is mathematical ([4], p. 52). The gist of this paragraph is, that *in modern science nature is revealed as mathematical in the original sense of "things as they can be taught and learned", i.e., things demonstrated and demonstrating themselves as congruent with our mental faculties.* This congruency becomes apparent in experimental science, where we ask questions through the instrumentality of our hands, and obtain answers fitting our mathematical design of nature, from which design the questions originated. The *mathesis universalis* may therefore be interpreted as the "fit between the hand and the brain"! Practiced by Newton, justified by Descartes, and metaphysically elaborated by Kant, this "fit" became the hallmark of modern science and the guarantee of scientific progress.

## CONSEQUENCES OF HEIDEGGER'S METAPHYSICS: SCIENCE UNLIMITED

Paraphrasing Heidegger's message, I conclude that science works because, in the mathematical design of nature, man discovered the fit between hand and brain. The growth of knowledge through experimental science spells the continued exploration of the universe along the lines of this fit. Science grows while feeling its way along the folds and wrinkles of this fit, by mapping out what may be called the *interface between mind and matter*. These metaphors, however, should be supplemented by an explicit statement of the metaphysics involved. Its main point is that the question of what matter or nature are "really" like, is rejected. There is no sense in conceiving of nature as something apart from or beyond our conceptualization, because all being depends on human existence. Loosely spoken, nature is an aspect of human existence, but since the birth of experimental science nature has acquired specific characteristics. Today, *nature is that aspect of human existence which emerges from the interaction of our intellect and our instrumentality (brain and hand!) as structured in the experimental natural sciences*. In summary, and using another metaphor: experimental science is a dialogue within human existence and nature is the content-matter of this dialogue.

It will be clear that in this view there are, in principle, no limits to the growth of scientific knowledge. As long as scientific inquiry continues, it will produce results. The metaphors used also speak of unlimited growth: maps can always be drawn in more detail and dialogues can continue endlessly.

This conclusion is in fact borne out by actual scientific developments. In the direction of the ever smaller, the ultimate elementary particle has so far retreated to ever receding horizons. In the other direction moreover, of the ever larger, the boundaries of the universe are similarly nowhere in sight. Yet, even if the limits of the universe in both directions should be demonstrated, whether theoretically or by observation, this would not necessarily impede the continued growth of science. In principle we could always find increasing complexity even within a finite universe. In this respect, recent developments in the biomedical sciences are illustrative. In fields like immunology and endocrinology, there is a steady growth of new data and of ever more complex patterns of interactions which in fact rest on a limited number of known units. The body fluids necessarily contain a limited number of peptides and protein molecules, the stoichiometric composition of which is well known. Yet there is apparently an uninhibited proliferation of proteinous and protein-related hormones, antibodies, blood-factors, pro- and anti-coagulants, etc. This multiplication of data is largely a process of uncovering new active configurations and fractions within the confines of a limited albeit very large pool of molecules. In fact, biological mechanisms at this level are unravelled by research which starts from unexplained or incompletely explained phenomena which themselves emerge from ongoing research. What happens is that looking at one and the same organism from ever-changeing perspectives, we continuously discern and define new phenomena which subsequently lead to new questions, hypotheses, and thus to research which discovers new mechanisms, factors, etc. Thus the unrestricted flexibility of our mental and experimental apparatus gives rise to continuous growth of scientific knowledge. This is especially clear in the biomedical sciences. Here life reveals itself to us in ever increasing detail and complexity as long as we apply our mental and observational capability of innovative definitions of life's phenomena.

## THE CASE OF MEDICINE

The previous section suggested metaphysical reasons for scientific growth as both progressive and unlimited. The biomedical sciences were a case in point. Does this conclusion also hold true for medical knowledge in general?

The simplest criterion for scientific progress was that science works, that it increases our grip on nature. In order to establish continued progress of medical knowledge, we should observe that medicine becomes ever more effective. Such claims should be secured by unceasing new therapeutical, palliative or preventive successes. For this claim I could of course invoke such evidence as transplantation surgery, new medicines, or *in vitro* fertilization. With equal ease, however, I could be refuted by contrasting evidence: the lung cancer epidemic, rampant resistance of micro-organisms against antibiotics, high rates of infant mortality in some countries, in others the increasing frequency of intractable diseases like Alzheimer's disease, multiple sclerosis, and many more. Considering the state of human health at population level, national as well as world-wide, it remains an open question whether modern medicine is indeed progressive in the sense of increasing efficacy. A commonplace answer to this problem is that medicine is defined by its diagnostic and therapeutic capabilities in individual doctor-patient interactions rather than by its efficacy at population-level. This reflects the contrast mentioned in section I, between medical science and the practice of health care. Let us now have a look at the philosophical implications of this definition.

It is usual to conceive of medicine as comprising the inventory of means and methods for preventing, diagnosing, treating, and alleviating the diseases and disabilities of human beings as well as the store of knowledge on which all this is based. It is also customary to further restrict the inventory to somatic interventions, excluding psychotherapy and social support. Medicine thus defined is considered a branch of the natural sciences with allied technologies. As such, medicine should have the potential for unlimited progress, as was already argued for aspects of biomedicine like endocrinology (section IV). At this point, however, we should realize that the inclusion of medicine in the natural sciences is problematic for several reasons. For instance, one may argue that every doctor-patient interaction has psychological and social dimensions which are not accounted for in the natural sciences. These dimensions influence medical interventions, whether or not the physician in question is fully aware of it. In fact, this is the reason why in medical experiments double-blind methods are necessary which are not required in biological or physical research.

Moreover, this pragmatic objection has a philosophical correlate. Having located natural science at the interface of mind and matter, one cannot use the same concept of science for interactions where two minds are involved. Medicine, other than biomedical science where *in vitro* material or animal models are used, works at the interface where subject meets subject. Herein lies the fundamental difference between natural science and the study of human beings. Let me explain.

Following Heidegger's interpretation, I argued in section III that nature was rendered scientifically knowable through the mathematical design of reality. This design was brought about by setting the "I", the knowing subject, against a reality rendered knowable through mathematics by divesting it of subjectivity. Natural science, in other words, precludes knowledge of subjects and this separates natural science not only from psychology and sociology, but also from clinical medicine, i.e., medicine as the study of *human* disease. The separation was already implicit in Descartes' metaphysical dualism. His *mathesis universalis* does by definition not relate to the contents of psychology. In Kantian terms, also, psychology could never become a true science. For Kant, the mind (*die Seele*) was not a phenomenon of nature.

Mark that this philosophical objection to the definition of medicine as a natural science does *not* rest on the well-known ethical grievance that science tends to "reduce" patients to organ systems, to neglect their human personhood, etc. It goes further than that: it is fundamentally impossible to obtain natural scientific knowledge about diseases (or related phenomena) as attributes of human subjects. In other words, all *medical* knowledge (other than knowledge of *in vitro* or other non-human models) *necessarily falls short of the criteria for natural scientific knowledge*. It cannot work with the efficacy, nor predict with the accuracy, of chemistry, physics, nor even biology. This conclusion is borne out when we compare medical theory with medical practice. Medical theory describes causes and mechanisms of many diseases. There is, however, no single disease where this knowledge suffices to explain *why and how this particular patient here and now shows these particular symptoms*. Neither does this knowledge suffice to predict how he or she will respond to specific medical interventions. But if medical knowledge cannot be defined as natural scientific knowledge, what other definition is required and feasible? To answer this question, I must retrace my epistemological argument from the post-Renaissance "scientific revolution" forward.

## THE GROWTH OF MEDICAL KNOWLEDGE, PAST AND FUTURE

The history of medicine is often described in pre-Kuhnian terms. Early roots are found in the Hippocratic tradition, step or jumpwise advance is seen through the ages, and the last 50 or 100 years are shown to bring crowning achievements and to ring in a glorious future. The suggestion of a unified body of knowledge evolving gradually throughout history is highly deceptive, however. Conflicting medical doctrines as well as divergent traditions of medical practice have always been rife (as ten Have has shown in his essay). Occupations like midwifery, surgery, and medicine were once wide apart and so were the knowledge and lore on which they rested. Until the 19th century there was no unified body of medical knowledge but only a variety of health-related disciplines of which academic medicine was one. The latter, moreover, had little impact on medical practice even after the conceptual shifts in medical theory in the course of the revolutionary 16th and 17th centuries. By way of example: both Harvey on circulation and Sanctorius on metabolism marked a revolutionary cognitive break-through. The involvement of anatomy and physiology in the "scientific revolution," however, did not entail much progress for medicine taken in a wider sense.

A major paradigm shift in medical science can, however, be claimed for the mid-19th-century. It was then that experimental medicine, founded by Claude Bernard and others, came under the spell of Descartes' *mathesis universalis*. As long as vitalistic principles were adhered to, living organisms were to some degree exempt from the laws of natural science. Vitalism was discarded by Bernard in his *Médecine Expérimentale* [2] as was Aristotelian space by Newton. To metabolism, laws of chemistry were applied just as two centuries earlier mechanical principles were applied to the heart. The work of Pasteur, Ehrlich, and many others contributed to the transformation of medical science into a natural science in the second half of the 19th-century. This vastly increased the practical implications of medical theory, which was one of the reasons why the various medical crafts were finally united into one profession. By 1880 this medical profession was established in all Western countries. It relied on academic medicine which was by now firmly rooted in natural science and had entered a most fertile phase of "normal science" (in Kuhnian terms) lasting until the present day.

Continuing in the tradition of late 19th-century medicine, modern medicine often still considers itself a branch of natural science. Some prudent reservations may be made to the effect that natural phenomena in human beings are less controllable and predictable than in lower organisms. That is why we need the *art* of medicine as a complement to medical science. Moreover, since the 1950s there is the trend, especially among psychiatrists and general practitioners, to consider medical science incomplete until fully integrated with medical psychology and sociology. It should be realized, however, that such amendments will not suffice. Philosophical reasons why medicine cannot be identified with natural science were offered in section V. For the same reasons, an effective integration between biomedicine and medical psychology and sociology is not to be expected. The improbability of an eventual unified medical science has been more fully discussed elsewhere [14]. Historically, we have witnessed several efforts of medical practitioners and theorists to break with the natural scientific definition of medicine. Prominent examples are the continental schools of *anthropological medicine* of the 1930s, and American *holism* since 1970. Both movements are discussed elsewhere [14]. Voicing discontent with the dominance of natural science, these movements strive for a redefinition of medicine on the level of individual doctor-patient interactions. During the past 20 years, another and differently oriented current has developed which questions the role of medicine at the macro-level: How does it function in relation to the health of the population? This is the view which considers scientific medicine an impediment to effective health care (section I).

From this brief survey I conclude that throughout history medicine has changed with regard to both the problems with which it was concerned and the type of knowledge on which it rested. During the past 100 years, it was mainly concerned with biological problems: bacteriological, biochemical, anatomical, and related causes and mechanisms of disease and disabilities. To these problems natural scientific knowledge was applied and this has led to growing effectiveness of medical knowledge in terms of predictability and controllability of the phenomena under study. In our days it is increasingly realized that those phenomena are not decisive for the problems which we today consider medically relevant. For instance, much is known about causes and mechanisms of infectious or metabolic diseases under strictly specified conditions. Using laboratory procedures or clinical experiments, scientific knowledge can be developed successfully in the tradition of the natural sciences. But under real life conditions this knowledge shows all kinds of practical shortcomings illustrated by the following examples. Finding the cause of legionnaires' disease enhanced the growth of medical knowledge. It has, however, not enabled us to predict when and where the ubiquitous *Legionella bacterium* is going to strike. Cancer research is now concentrating on molecular analysis of the mechanism of malignant degeneration of cells, and is making rapid progress. It is no help, however, in spotting the environmental factors conducive to the occurrence of specific cancers. From a natural scientific point of view, tuberculosis is a vanquished disease, but preventive measures and bacteriostatic medicines are ineffective against tuberculosis when inadequate housing and dietary deficiencies maintain high rates of (re-)infection and low levels of physical resistance.

All this serves to show that the case of medicine is a special one, indeed. As long as medical problems are framed in terms of biological models, controlled experiments or well-specified clinical situations, they can be solved by natural scientific methods. Medical knowledge thus obtained is effective and grows progressively by natural scientific standards. Epistemologically it can be explained, I believe, in Heidegger's terms and is expected to show unlimited future growth. At the same time, however, we are confronted with the inadequacy of the problem definitions mentioned above.

The need for a redefinition of medical problems, and thus of medicine itself, is felt at two fronts. First, in individual medical practice attention is claimed for the personal experiences of patients and physicians. This refers not only to the psychological dimensions of illness and medical interventions. Questions about *personal responsibility in health and disease* are increasingly being asked. What influence does an individual have on his own health? Given his personal life-style, will he get this or that disease and will it be fatal in his case? Such questions will not be appeased by the statistical answers offered by science. And secondly, at the population level, medicine faces problems of social equality and effectiveness. Can malaria be eradicated, can blindness by vitamin A deficiency be prevented, will it be possible to turn the tide of the threatening AIDS epidemic? These problems are not solved by mere knowledge of preventive and therapeutic procedures. On both fronts the definition of medicine as a branch of natural science proves *inadequate* and medical knowledge shows as yet little growth or progress.

Yet such knowledge as is available is not static – witness the ongoing discussions about health care systems, about priorities in care and research, and about "the role of medicine" [9], as well as the growing interest in medical ethics, and in the psycho-social and cultural aspects of medicine [12]. In these discussions medical knowledge is being redefined as *all knowledge that is relevant for solving health-related problems in their real-life setting*, and this setting comprises the full spectrum of psychological, sociological, cultural, and economic dimensions of health and disease. Is medical knowledge thus redefined progressive? How far is future growth to be expected and by what criteria should it be evaluated? My answers to these questions are brief and inconclusive. Moreover, they are of an *anthropological* rather than an *epistemological* nature. Let me explain.

## CONCLUSIONS IN ANTHROPOLOGICAL PERSPECTIVE

It is indisputable that science and technology, especially in the last 100 years, have become immensely powerful. This has markedly changed all aspects of our culture, medicine included. Modern science and technology are continuously recreating the world and our lives. This "re-creation" should be taken literally. Man and world are newly created because scientific achievements are not merely instrumental in an external sense. Referring to our earlier epistemological discussion, we may now say that science and technology act essentially from "inside" reality as defined at the mind-matter interface. Thus, science truly and operationally *recreate*s humanity.

Humanity is what medicine is about. So if natural science were fully constitutive of medicine, human health and disease might be recreated as a product of experimental research and technological support. This is in fact what was feared and resented some 30 years ago. In many Western countries these fears led to a cultural revolution with anti-scientific tendencies. But that was long ago, and since then much has happened in medicine and in Western culture. For one thing, the notion of medicine as more than natural science has taken root within at least some sub-groups of the profession as well as among the public. Efforts have been and still are being made at comprehensive therapeutic approaches and at interdisciplinary medical research. It has also become apparent, however, that such efforts are of little avail; or, rather, that their effectiveness can hardly be proven in any objective way. This demonstrates the ambiguity of our present position. We clearly do not accept a vision of mankind as an object that can only be *measured* and manipulated. Yet we cannot seem to escape employing objective methods of measurement. Many similar observations may be made. There is a general abhorrence of manipulative medicine, of machines taking over organ functions, and computers dominating human decisions. Yet new technical supports, as they become safer, cheaper, and often smaller, are tacitly accepted by increasing numbers of consumers. What was once an unnatural monstrosity, gradually emerges unobserved into the cultural foreground.

This shows that mankind is changing, indeed, albeit that the changes are gradual and may span an entire generation. Older members of the population may still defend values which are hardly noticed by the next generation. The changeability of mankind and of human values cannot be elaborated here ([14], pp. 148 ff.). It is, however, important to evaluate changes where scientific measurement *fail*s to achieve progress. A popular (because democratic) answer is to equate success with cultural acceptance. In the case of medical knowledge this implies that the biomedical achievements of the late 19th and early 20th centuries did indeed signal progress, since the general attitude toward science in those days was one of "great expectations". Similarly, the present expansion of alternative types of medicine should today be considered progressive because of its public recognition. But major technical advances like *in vitro* fertilization and embryo transfer, because they are still hotly disputed by major sections of the population, cannot today be called "progress", though they may become progressive tomorrow!

Is this too easy an answer? There seems to be no single, acceptable alternative. It is clear that what we are looking for are new norms. "Humanity" has often been invoked as a criterion, but more detailed and specific descriptions of this ideal have not been forthcoming. Religious systems still inspire many citizens, but when addressing medical issues, people are often internally divided. For a more comprehensive analysis of moral values in modern culture I must again default. I simply refer to the analysis of A. MacIntyre [8]. Anyway, *today* there is no stable basis for shared values by which we may all judge medical advances.

The changeability of mankind, however, which led us into this blind alley, should still make us hopeful. There is no reason to despair of achieving a consensus and eventually stability. This hope places a heavy responsibility on all who are personally inspired by values that they consider beneficial for mankind. To sustain and propagate virtues is of special urgency in an age that is "after virtue". What goes against the grain today may inspire a revolution tomorrow. In a time when a cultural plebiscite determines what constitutes progress, medical or otherwise, it may be noble to defend what the vast majority rejects.

*University of Leiden,*
*Leiden, The Netherlands*

## BIBLIOGRAPHY

1. Bernal, J.D.: 1965, *Science in History*, Watts & Co, London.
2. Bernard, C.: 1865, *Introduction à l'étude de la Médecine Expérimentale*, Baillière, Paris.
3. Bryant, J.: 1969, *Health and the Developing World*, Cornell University Press, Ithaca and London.
4. Heidegger, M.: 1962, *Die Frage nach dem Ding*, Niemeyer, Tübingen, F.R.G.
5. King, M.: 1967, *Medical Care in Developing Countries*, Oxford University Press, Nairobi and London.
6. Kuhn, T.S.: 1970, *The Structure of Scientific Revolutions*, University of Chicago Press, Chicago and London (especially Ch. XIII and Postscript).
7. Kuhn, T.S.: 1977, *The Essential Tension*, University of Chicago Press, Chicago and London.
8. MacIntyre, A.: 1984, *After Virtue*, University of Notre Dame Press, Notre Dame, Indiana.
9. McKeown, T.: 1984, *The Role of Medicine*, Blackwell, Oxford.
10. Medawar, P.B.: 1974, *The Hope of Progress*, Wildwood House, London.
11. Niiniluota, I.: 1984, *Is Science Progressive?*, D. Reidel Publishing Co., Dordrecht, Netherlands.
12. Payer, L.: 1988, *Medicine and Culture*, Henry Holt and Company, New York.
13. Popper, K.: 1973, *Objective Knowledge*, Clarendon Press, Oxford.
14. Steen, W.J.van der, and Thung, P.J.: 1988, *Faces of Medicine*, Kluwer Academic Publishers, Dordrecht, Netherlands (especially Ch. V, VI).

B. INGEMAR B. LINDAHL

# THE DEVELOPMENT OF POPULATION RESEARCH ON CAUSES OF DEATH: GROWTH OF KNOWLEDGE OR ACCUMULATION OF DATA?

## INTRODUCTION

The great importance of cause-of-death data for historic, medical and social science research is undisputed. For a long time in the early history of public health planning in Europe, as well as in other parts of the world, information on causes of death were the only nationwide data available on the health of the population.[1] Although other measures were also used, such as infant mortality rate and mean life expectancy, the cause-of-death statistics were the only data on specific conditions. Even so, the occurrence of fatal conditions is a very limited measure of the state of health in a population. Nowadays we have access to morbidity data both through registers and statistics. Yet the cause-of-death data have not outlived their usefulness. They are still an important source of information in themselves and a valuable complement to morbidity data in population research and public health planning.

A study of the nineteenth and twentieth centuries international development of cause-of-death registration and statistical tabulation clearly shows that the amount of data on causes of death has gradually increased. It is not only a matter of a total increase, but also an increase in proportion to the number of deaths, and of data on the individual cases. Based on these cause-of-death data, a number of theories have been constructed in population research. With insight on how the cause-of-death data are usually collected one may question, however, to what extent, if any, the resulting increase of cause-of-death theories amounts to, or has led to, an *improvement* in knowledge of causes of death. One may ask, to what extent, if any, there has been a growth of scientific knowledge in population research on causes of death, and to what extent merely an accumulation of data. As will be emphasized in our analysis, the answer depends on the *relevance* of the cause-of-death data on which the theories are based, to the scientific problems the theories are constructed to solve; a mere increase of data, or of theories, even an improvement of the closeness to truth of data or theories, is not sufficient for a growth of knowledge. The relevance of these data and theories to the problem to be solved is a *sine qua non* for growth of knowledge.



*H.A.M.J. ten Have et al. (eds.), The Growth of Medical Knowledge, 103–119.*

Originally the authorities' routine collection of cause-of-death data was meant for official statistics purposes only, i.e., public health planning, and not for general scientific use.[2] Despite this, these data have been used to an increasing extent also in scientific research.

In historic demography and history of medicine research on the eighteenth and nineteenth centuries, the official *statistics* itself is still often the only source available. For these periods primary cause-of-death data on individuals exists in Europe, generally only regarding larger cities. In social science and medical research on causes of deaths in contemporary society, cause-of-death *registers* and *death certificates* are used as primary sources of information. As a result, conjectures are made, some of which are less advanced, what we may call "hypotheses", and some more advanced, which we may call "scientific theories". The former are primarily derived from reading and interpreting official statistics, while the latter are based on primary data and on the official cause-of-death register, often linked with other registers containing information on, e.g., in-patient diagnoses, socio-economic status, etc. The nature of these more advanced conjectures, and the justification for calling them "scientific theories", are discussed in the second part of this essay.

Since the turn of the century, great international efforts have been made, and are still made, to increase the amount and improve the quality and comparability of cause-of-death data. Considerable improvement has been achieved in the data collection process. These will be briefly accounted for in the third part.

A number of important theoretical and methodological problems remains to be solved, however. One of the most serious problems, according to Preston *et al.* [25], affecting the international comparability of cause-of-death data, is the *selection of causes* when more than one disease, injury or other condition has been identified. A first selection is made by the certifier (the physician) when reporting the causes – today usually on a death certificate. Since in most countries not all causes on the certificate are registered, a second selection is often made by the registrars when registering the causes from the death certificates at the statistical offices. The principles for making these selections vary between nations, and the international agreements reached have changed several times during the twentieth century [17, 21]. In many countries the statistical offices register only *one* condition from each death, but also in those nations where more than one condition is registered, a principal cause is pointed out and recorded separately. Therefore, when dealing with the problem of selecting causes of deaths on national and international levels, attention has been focused on the selection of the principal cause of death. The problems are just as great, however, when selecting the other conditions, i.e., the consequences of this principal cause and the otherwise contributory conditions. Basically, the purpose of the selection, the scientific problem to be solved in the study, determines how many, and which, causes are relevant to select from an individual death. These problems have been analyzed and discussed in previous studies [14, 15, 16, 17, 21]. The results of these studies, on which our analysis will be based, are presented in the fourth part.

In this essay we shall confine our analysis to investigating to what extent, if any, it is possible to learn of the relevance of cause-of-death theories to the problems they are constructed to solve, i.e. to test their relevance. This relevance is one criterion of the scientific value of cause-of-death theories, their testability is another. The purpose of this essay is to establish, at least tentatively, the scientific value of cause-of-death theories in population research, by examining their testability in this respect.

## THE NATURE OF CAUSE-OF-DEATH THEORIES

Although scientific knowledge may consist in, or develop through, more than only hypotheses and theories, contemporary philosophy of science discussions and analyses concerning the growth of scientific knowledge focus to a large extent on the advancement of hypotheses and theories.[3] Accordingly, in this essay our discussion of the possibilities of a growth of knowledge will be restricted to the scientific value of cause-of-death *theories*.

The reason I choose to call the conjectures, to be discussed in this essay, "theories" and not "hypotheses" is to indicate that they are not merely first guesses but the result of studies carried out with conventional scientific methods.[4] These theories are the result of, e.g., historic, demographic, epidemiologic, and social science research. They are not merely speculations from, e.g., reading and interpreting official statistics. However, the degree of sophistication of these conjectures, and the distinction between 'hypotheses' and 'theories', are not crucial of the discussion in this essay. Even more simple statements, based only on reading and interpreting cause-of-death statistics, are affected by the analysis and conclusions of this essay.

The forms in which the cause-of-death theories are expressed and the methods by which they have been reached, e.g., the types of measures of occurrence used, the types of comparisons made, and the study designs applied, are not important for our analysis. The crucial point is the relevance of the data on which the theories are based. To make our analysis more concrete it may be pointed out, however, that the type of conjectures referred to here as "cause-of-death theories" are empirical generalizations of probabilistic character. The theories may be expressed in terms of a relative risk. An example in the most simple form is "The risk of *A* is greater given *B* than given not-*B*", where "*A*" stands for death and "*B*" for a disease, an injury, or external (exosomatic) cause of an injury. (In this case the purpose is not to compare the causal influence of *B* with the causal influence of any other particular cause, but only with the causal influence of the absence of *B*.) Another example is "The risk of *A* is greater given *B* than given *C*", where "*A*" and "*B*" have the same meaning as in the previous example and "*C*" refers to another disease, injury, of external cause of an injury than "*B*" does. In both these examples the immediate purpose of constructing the theory is to predict death for a certain category of people. Other purposes, to be discussed later on, require other types of theories.

It is a controversial issue in philosophy of science whether scientific theories can have truth values, i.e., be true or false, and even whether they can at all be more or less close to truth, or have, what Popper calls, "verisimilitude".[5] Without taking sides on this issue regarding more sophisticated theories, within, e.g., physics, it will be assumed in our analysis of cause-of-death theories that these comparatively simple theories can be true or false or be more or less close to truth.

When talking of truth and falseness (and closeness to truth) of cause-of-death theories we will employ a realistic view of truth, i.e., that the truth or falseness (or the closeness to truth) of data and theories depends on their correspondence or lack of correspondence (or degree of correspondence) with facts.[6] The ontological status of the theories is not decisive for our analysis. It seems appropriate, however, to confine our analysis to theories of objective knowledge, in Popper's sense of "objective knowledge", i.e., theories "published in journals and books and stored in libraries ..." ([24], p.73).[7]

## THE DEVELOPMENT OF CAUSE-OF-DEATH REGISTRATION

As we have already noted, the cause-of-death theories in population research are based on routinely collected data from death certificates, official registers and statistics. A brief account should here be made of the most important of the earlier mentioned international efforts made to increase and improve these data.[8]

The problem that has attracted greatest interest in recent studies on the validity of cause-of-death data is the question of empirical evidence supporting the diagnoses. Often data in one source, e.g., on death certificates, are compared with data in another, e.g., clinical records or autopsy reports; also diagnoses based on one type of investigation, e.g., pre-mortem examination, are compared with diagnoses based on another, e.g., autopsy [9].[9] In order to improve validity, in this respect, better medical competence, i.e., certification of causes of death to a greater extent by physicians, and an increased frequency of autopsies, are called for. To different extent these measures have already been taken in many countries. In some countries the autopsy rate has declined during the last two decades, however.

With the development of the *International Classification of Diseases, Injuries, and Causes of Death* (ICD) the number of conditions and events possible to register has increased. The revisions of the classification have also resulted in a specification and division of the categories.

Although the classification is not intended to be a nomenclature, it has *de facto* led to an increased standardization of terminology and to greater possibilities of translating and comparing data between languages and populations.

The death certificate form, introduced on an international level by WHO in connection with the sixth revision of the ICD in 1948, enabled the certifiers in many countries to record more conditions on each individual than had previously been possible. It became possible to specify the conditions through which the principal cause had caused death and to record other contributory conditions.

With the sixth revision of ICD, international definitions of basic concepts, like 'causes of death', 'underlying cause of death', and 'contributory conditions', and rules for selecting and classifying these events were introduced. International instructions for physicians [35] and statistical registrars ([34], pp. 345–352) were published.

All these actions have contributed in different ways to increase the amount and to improve the quality of data and consequently of cause-of-death theories.

## THE SCIENTIFIC VALUE OF CAUSE-OF-DEATH THEORIES

The scientific value of cause-of-death theories is determined by (i) their truth value (or closeness to truth), (ii) their relevance to the problem to be solved, and (iii) their testability with regard to (i) and (ii). This means that a theory

$T_1$ is to be preferred to another theory $T_2$, if $T_1$ is closer to *relevant truth* than $T_2$.[10]

Thus, in order to test the scientific value of a cause-of-death theory we must be able to establish not only (a) whether or not (or to what extent) the theory is true, but also (b) whether or not (or to what extent) the theory is relevant to the problem to be solved.

The first is a large enough problem in itself. It is a matter of both the validity of basic data, often measured, as we have seen, in terms of, e.g., the certifiers' medical competence and the diagnostic methods used,[11] and also of the accuracy of the statistical calculations made. Although germane to the question of the testability of cause-of-death theories, none of these issues will be dealt with in this essay. Nor will we deal with the epistemological issue, whether or not (or to what extent) it is possible for us *to know* if data, and consequently a theory, is true or not (or more or less close to truth). Instead, we shall concentrate on the second problem, the relevance of theories to problems.

### *The Relevance of Theories to Problems*

The relevance of theories to problems may, of course, be improved by modifying either the theory or the research problem. We shall, however, start from the assumption that the research problem is fixed and that it is the theories that shall be constructed to be relevant. The relevance of a cause-of-death theory to a research problem depends on the selection criteria and on the purpose(s) for which the causes of death have been selected.

The purposes of cause-of-death studies vary considerably. I have discerned elsewhere [17] five purposes of causal accounts, commonly referred to in causal theory in general: (i) to *describe* what caused death; (ii) to *explain* why death occurred; (iii) to identify causes, the manipulation of which would *prevent* untimely death; (iv) to identify risk factors, which make it possible to *predict* untimely death; and (v) to *adjudge personal responsibility* for death. The concepts 'describe' and 'explain' are here somewhat narrowly defined.[12]

All these purposes are more or less prevalent in all disciplines concerned with cause-of-death research. A theory may also be constructed to serve more than one of these purposes.

It may seem reasonable to assume that all medical studies have a common *ultimate* purpose, to contribute to the elimination or at least minimization of human suffering and of diseases, impairments, disabilities, etc., and also of untimely death. Yet far from all medical investigations have treatment or prevention as their *immediate* purpose. Only a fraction of cause-of-death studies aim at acquiring knowledge possible to put to practical use. The immediate purposes of cause-of-death studies may be to merely explain or calculate the risks of untimely death. These explanations and calculations can then be used in further studies, in order to find the causes suitable for treatment and prevention. A single study may, of course, have more than one immediate purpose, e.g., both to explain untimely deaths and to identify causes, the treatment or prevention of which would reduce such untimely deaths in the future. However, in order to make as clear as possible the points to be made in our analysis, we shall focus on studies with only *one* immediate purpose.

*Description* is often the immediate purpose of historic demography and history of medicine research. The problem may consist in the lack of exact and complete data about the distribution and frequency of causes of death in general in a certain region at a specific time. Relevant would be any true and non-biased theory about the frequency of causes of death in this region at this time.

*Explanation* is probably a more common immediate purpose of historic research. This is also often the primary purpose of epidemiologic research on causes of death occurring in populations of today. The problem may be an observed excess mortality or a significant trend in the mortality which causes are insufficiently known. For a theory to explain these observations it must point out causes that, e.g., are not normally occurring in this context or which otherwise are unknown to those requiring the explanation.

*Prevention* of untimely death is often the immediate purpose of research meant to provide a basis for immediate action. The problem may be for example an excess mortality in a certain population. For a theory to be relevant it must point out causes, the elimination or changing of which would eliminate or at least reduce the mortality in this population. In contradistinction to theories for an explanatory purpose, the theories for a preventive purpose need not refer to unusual or unknown causes.

*Prediction* is perhaps the most common immediate purpose of epidemiology. The problem is often that the fatality of a certain disease, injury or external cause of injury is insufficiently known. A theory indicating the seriousness of these causes, in terms of, e.g., causal efficacy or sufficiency, would then be relevant.

*Adjudging personal responsibility* is the immediate purpose of some research in epidemiology and treatment evaluation. Classical examples are studies on suicide and homocide. The problem may also be, e.g., that an inadequacy in a common surgical treatment has been detected and it is insufficiently known to what extent this inadequacy causes death. A theory indicating the frequency of deaths attributed to the action in question in this type of population would then be relevant.

In this brief and sketchy overview of prevalent purposes and problems only a few criteria of relevance of theories have been mentioned. Many more criteria could be accounted for.[13] Implicit is also the important fact that different theories can be relevant to a problem in greater or lesser degree.

So far we have only discussed how cause-of-death theories *ought* to be constructed in order to satisfy prevalent needs and interests. We shall now proceed to examine the basis of how cause-of-death theories are *actually* constructed.

### *The Actual Data Collection Process*

In previous studies several theoretical and methodological problems in the cause-of-death data collection process have been pointed out and analyzed [15–17, 21]. These deficiencies in the data collection process make it difficult, sometimes impossible, to test the relevance of theories based on these data.

One problem discussed is the insufficient concordance between purpose and method in the registering of the principal cause of death, the so called "underlying cause of death". 'The underlying cause of death' is defined as "(a) the disease or injury which initiated the train of events leading directly to death, or (b) the circumstances of the accident or violence which produced the fatal injury" ([36], p. 763). The WHO's purpose of defining the principal cause of death as a cause that initiates the course of events in this way is that this point is supposed to be the most suitable to act upon in order to prevent untimely death.

> "From the standpoint of *prevention of deaths*, it is important to cut the chain of events or institute the cure at some point. The most effective public health objective is to prevent the precipitating cause from operating" (my italics [36], pp. 699–700).[14]

It has been pointed out that whereas the purpose is prevention, and therefore requires, e.g., irreplaceable and manipulable causes to be selected, some of WHO's guidelines[15] for selection of the underlying cause give preference to more serious conditions over less serious and thus serve a purpose of prediction rather than prevention. None of WHO's rules gives preferences to causes especially suitable for treatment or this type of prevention [15, 17].

Another problem is that the definition of the 'underlying cause of death' gives no guidance in the selection between joint and parallel causes of death [15, 17, 21].

A third problem is the manifold meanings of the definition of 'contributory conditions': "any other significant condition which unfavourably influenced the course of the morbid process, and thus contributed to the fatal outcome, but which was *not related to* the disease or condition directly causing death" ([36], p. 700; my italics). The question here is how "not related to" shall be understood. A closer examination of this definition indicates that it seems to allow only causally redundant conditions, without significant explanatory, predictive or preventive value, to be entered as contributory conditions on death certificates ([17], p. 145).

Due to these and other problems in the instructions for physicians and statistical registrars, it is difficult to know how causes of death *should* be recorded, according to WHO, and, since physicians often are insufficiently informed of these instructions and seldomly motivated enough to carefully follow them, it is even more difficult to know how, and for what purpose(s), the data have *actually* been recorded.[16] In this respect the situation is not much different from the registering of causes of death in the eighteenth and nineteenth-centuries, when there were no international instructions (and in many countries not even national instructions). Without this knowledge, we cannot know if the theories based on these data are relevant to the problems the studies are set out to solve; we cannot test the relevance of the theories.

### *An Example*

As an illustration of the fact, that if we do not know how data have been selected we cannot establish their relevance to the problem we have set out to solve, consider the following example from a study made a few years ago [14].

The point of departure was an observed *threefold mortality increase* for both men and women attributed to rheumatoid arthritis (RA) as the underlying cause of death in the official statistics for Sweden between 1971 and 1975. All death certificates for 1971 and 1975 with RA registered by the National Central Bureau of Statistics (NCBS)[17] as either underlying cause or complication/contributory condition were studied. Physicians were found to have reported a slight *decrease* for men and practically *no change* at all for women between the years.

Disregarding the fact that the register data are based on the death certificates, this could just as well have been the result of two separate calculations on the risk to die from RA, one based on the NCBS cause-of-death register, the other on physicians' statements on the death certificates. Which would then have been most correct?

Again, the question whether or not (or to what extent) these theories are true, will be put aside here. Let us assume that we know that both theories are true, in the sense that all persons died from a sequence of causes involving RA and the other conditions recorded. Our question is, whether or not (or to what extent) these theories are relevant to the problem to be solved. How do we answer this latter question?

If we start from the explicit WHO purpose of studying causes of death, to prevent untimely death, the criterion for testing the relevance of these two theories would be their *preventive value*. The question is then: What would have been the most effective means to prevent untimely death: (i) preventing the operation of the causes selected by the physicians as the underlying cause, or (ii) preventing the operation of those selected by the NCBS? We must also know if it were even feasible "to cut the chain of events or institute the cure" at these points. These questions cannot be answered by merely studying the underlying cause-of-death *diagnoses* (the statements on the certificate) – at least not in isolation. Nor does it suffice to use our knowledge of cases *in general*, i.e., how effectively untimely death *usually* can be prevented by preventing causes of this kind from operating, or how feasible it *usually* is "to cut the chain of events or institute the cure" at these points. We need to know the facts of the *particular* case, and only the certifying physician has this knowledge.

Thus, to sum up, the testing of the relevance of cause-of-death theories, as defined in this essay, at least requires a better defined purpose of collecting the underlying cause-of-death data and also more knowledge about the *actual* basis of the certifiers' selection of the underlying cause of death. The same thing goes for the multiple cause-of-death theories, which include also complications and contributory conditions, only the problems are much greater.

## CONCLUDING REMARKS

In order to do full justice to the usefulness of cause-of-death data from death certificates, official registers, and statistics, a more thorough analysis than we have been able to make in this essay is necessary. However, a few conclusions are possible to draw from our analysis.

The first and major justifiable conclusion is that the knowledge we have, of how and for what purpose(s) the routinely collected cause-of-death data have *actually* been selected and registered on death certificates and in official registers and statistics, is not sufficient for us to be able to establish the relevance of cause-of-death theories to the problems they are constructed to solve. The cause-of-death theories cannot be tested in this respect. Consequently, the lack of this knowledge makes the scientific value of cause-of-death theories, if not completely non-existent, at least considerably impaired.

The second conclusion, of which we can be equally sure, is that, due to the lack of this knowledge, we are not able *to know* to what extent, if any, there has been a general growth of scientific knowledge of causes of death in population research. "General growth" here refers to an improvement in all the three respects, (i)–(iii), mentioned at the beginning of above section, and regarding all, or most, of the causes registered in one nation in one period compared to another.

The third conclusion, of which we cannot be sure, but anyway have good reasons to make, is that there has not *in fact* been a general growth of scientific knowledge of causes of death in population research. From what is known of the way routinely collected cause-of-death data are registered, and how they are used in population research, it seems like there has been more of an accumulation of data than a growth of scientific knowledge in the development of this field.

Finally, we may also conclude that a new kind of validity studies of cause-of-death data is needed, that examines not only the truth but also the relevance of these data. A first step in this direction is to further analyze the purpose, or purposes, of registering causes of death.

*Karolinska Institute,*
*Department of Social Medicine,*
*Huddinge University Hospital, Sweden*

ACKNOWLEDGEMENT

I wish to thank Filip Cassel and Lars Elffors for their helpful criticism of earlier drafts of this essay, and Alan McMillion for improving my English.

NOTES

[1] Among the earliest efforts to tabulate all deaths in the population by cause were the

statistics commencing in Sweden (1749), England and Wales (1861), Italy (1881), New Zealand (1881), Japan (1899), and USA (1900) ([25], regarding Sweden, see [17, 21, 22]).

Examples of how cause-of-death data were used in public health planning are the famous studies in Britain by William Farr [7, 11]. In studies published in 1839, i.e., even before the nationwide registration of causes-of-death, Farr compared the mortality and causes of death in urban and rural areas. The results of one of these, comprising a rural population in the selection brought down to the level of the urban (3.5 millions), showed that the mortality in all diseases was greater in the cities, but also that the fatality from zymotic diseases and diseases of the respiratory organs were about twice as common in the cities. These and other findings led to recommendations for sanitary reforms ([7], pp. 129–130).

[2] The purpose of the earliest cause-of-death statistics, for example in Sweden, was to serve the economic calculations of the state, and not primarily humanitarian and medical research needs ([27], pp. 227–229). A major vital statistics problem before the nineteenth-century in countries like Britain, as well as France and other nations on the continent, was the absence of a census taken at regular intervals. Eyler points out that a reason for the reluctance in eighteenth-century England to take or especially to publish a census was "the feeling, born no doubt of mercantilist theory, that population figures were crucial state secrets" ([7], p. 39).

[3] See for example the discussion in [29]. Examples of a theory being superseded by another theory, in Popper's sense of 'superseded', is given below, n. 5.

[4] Rapoport [26] distinguishes between four senses of 'theory': (1) "a collection of derived theorems tested in the process of predicting events from observed conditions" (used in natural science, e.g., physics and mathematical genetics); (2) "definitions" of, e.g., 'social action' made through "selecting events and combining them in such a fashion as to make the terms applied to these combinations fruitful" (used in, e.g., social science); (3) "metaphors, concepts, and definitions" intended "to achieve and to impart intuitive understanding of social behavior, of the nature of institutions, of political systems, of cultures, and such matters" (used in, e.g., social science and depth psychology); and (4) "the normative, value-laden sense" (used in, e.g., political science "concerned, for example, with the question of what is the best form of government"). (Rapoport's second and third sense of 'theory' seem to overlap. Both include definitions.)

In analyzing this taxonomy of theories Suppe [29] discernes additional interpretations of 'theory' in Rapoport's text. Instead of four, Suppe finds seven meanings of 'theory'. One of these seems to include what is called in the present essay "cause-of-death theories", namely "stochastic theories, which are mathematical and quantitized but the basis for quantitizing is counting rather than measuring; such theories generally do not involve a notion of state, rather employing the notion of an event; characteristic problems are sampling strategies and statistical validation; paradigm examples include theories in genetics, demography, epidemiology, and ecology" ([29], p. 123).

[5] It lies outside the scope of this essay to discuss whether cause-of-death theories can be true or false or only more or less close to truth or have what Popper calls "verisimilitude". In order to indicate that this question is left open, I will use

expressions like "the truth or falseness (or the closeness to truth)", "truth value (or closeness to truth)", "whether or not (or to what extent) the theory is true", etc.

In *Conjectures and Refutations* Popper accounts for his idea of a theory being a better or worse approximation to truth (i.e., his concept 'verisimilitude'). Popper gives a list of types of cases that clarifies his concept. A longer citation is therefore motivated:

"I shall give here a somewhat unsystematic list of six types of cases in which we should be inclined to say of a theory $t_1$ that it is superseded by [another later theory] $t_2$ in the sense that $t_2$ seems – as far as we know – to correspond better to the facts than $t_1$, in some sense or other.

(1) $t_2$ makes more precise assertions than $t_1$, and these more precise assertions stand up to more precise tests.

(2) $t_2$ takes account of, and explains, more facts than $t_1$ (which will include for example the above case that, other things being equal, $t_2$'s assertions are more precise).

(3) $t_2$ describes, or explains, the fact in more detail than $t_1$.

(4) $t_2$ has passed tests which $t_1$ has failed to pass.

(5) $t_2$ has suggested new experimental tests, not considered before $t_2$ was designed (and not suggested by $t_1$, and perhaps not even applicable to $t_1$; and $t_2$ has passed these tests.

(6) $t_2$ has unified or connected various hitherto unrelated problems.

If we reflect upon this list, then we can see that the *contents* of the theories $t_1$ and $t_2$ play an important role in it. (It will be remembered that the *logical content* of a statement or a theory $a$ is the class of all statements which follow logically from $a$, while I have defined the *empirical content* of $a$ as the class of all basic statements which contradict $a$.) For in our list of six cases, the empirical content of theory $t_2$ exceeds that of theory $t_1$.

This suggests that we combine here the ideas of truth and of content into one – the idea of a degree of better (or worse) correspondence to truth or of greater (or less) likeness or similarity to truth; or to use a term already mentioned above (in contradistinction to probability) the idea of (degrees of) *verisimilitude*" ([23], pp. 232–233; italicized in the original).

[6] The question of whether or not a realistic view of truth applies also to more sophisticated theories than the cause-of-death theories is left open here.

[7] Popper distinguishes between "subjective knowledge" and "objective knowledge, or knowledge in the objective sense, which consists of the logical content of our theories, conjectures, guesses (and, if we like, of the logical content of our genetic code)" ([24], p. 73).

[8] This historical development is accounted for in greater detail in [17, 21].

[9] It may be questioned whether these studies really validate the cause-of-death diagnoses. The studies only show whether the diagnoses also occur in other medical documents and what methods of investigation have been used and what medical competence the examining physician had. This could possibly be looked upon as a probabilistic way of measuring validity, namely the probability – although not necessarily numerically defined – that the examining physician and/or the methods used have the capacity of making correct diagnoses within the field under study.

For an overview of cause-of-death validation studies, see [1, 9].

[10] This position is inspired by Popper's view that "the fittest hypothesis is the one which best solves the *problem* it was designed to solve, and which resists criticism better than competing hypotheses" ([24], p. 264; italicized in the original). According to Popper, the trying out and discarding of theories is done "by trying to get nearer to the truth – to a fuller, a more complete, a more interesting, logically stronger and more *relevant* truth – to truth relevant to our problems" ([24], p. 148; my italics).

The application of Popper's philosophy of science to epidemiologic research has been discussed in the epidemiologic literature [2–5, 12, 28] reprinted in [10]; see also [8, 13, 18–20], [31–33] and ([30], pp. 82–93). However, none of these studies discuss the relevance requirement or the difficulties of testing the relevance of data and theories.

[11] See note 9.

[12] It may be in accordance with common usage to say that *all* accounts are descriptions and that the accounts of the types (ii)–(v) should therefore be viewed as subcategories of (i). It may also seem natural from an everyday language point of view to say that all *causal* accounts are explanations, since they tell us why things happened. However, in the analysis of cause-of-death accounts, i.e., on death certificates, in registers, and in the form of theories, it is more fruitful to define 'description' and 'explanation' more narrowly [17].

In this essay 'description' is restricted to a category excluding accounts for explanatory, preventive, predictive and adjudging personal responsibility purposes. An example of a description in this sense is an account of causally redundant causes in cases of overdetermination (i.e., when the effect was brought about by more parallel causes than was sufficient for the effect) when the other causes, in themselves sufficient for the effect, is already known to those acquiring the data, e.g., those making the study. This seems to be the idea behind the WHO's concept 'contributory conditions' ([17], pp. 142–145). The purpose of seeking information about causally redundant causes can be a striving for a complete account of *all* causes involved.

For a causal account to be an explanation, in the sense used in this essay, it needs to point out causes unknown to the persons acquiring the data. This is not necessary for accounts made for the other four purposes.

[13] For a fuller account of purposes and criteria of selecting causes of death, see [17].

[14] WHO uses here "precipitating cause" as synonymous with "underlying cause".

Unfortunately, WHO does not further indicate why this initiating cause should always be the most effective point at which to prevent untimely death. It is not easy to see why, or in what way, the position in the chain of events determines the possibilities to treat or eliminate the cause and/or determines, e.g., its degree of irreplaceability. (For further analysis of the criteria suitable for the purpose of prevention, see [17]).

[15] The guidelines are Rule 6 and two of the *Notes for use in underlying cause mortality coding* ([36], pp. 707–708, 713, 718). (See [15], pp. 146–147.)

[16] As far as I know, no empirical study has been made of how, and according to what criteria and for what purposes, physicians *actually* register, i.e., establish and select, causes of death. In conventional validity studies, *establishing* causes of death (i.e., the finding out of whether or not the person had the condition in question and died from

it) is often not distinguished from the *selection* of already established causes (see, e.g., the studies accounted for in [1] and [9]). These two problems are usually lumped together and simply spoken of as "establishing causes of death" or "making cause-of-death diagnoses", etc. Similarly, physicians are probably not always aware of the selection process when filling out death certificates.

The physicians' actual purposes when identifying (i.e., establishing and selecting) causes of death is probably often the same as the purposes for making diagnosis in living patients. Elffors identifies five aims for making diagnosis in clinical practice: "(i) to obtain a guideline for the choice of therapy; (ii) to obtain a tool for understanding the clinical course and making of prognosis; (iii) to perform a primary scientific act; (iv) to perform a formal/medico-legal act; and (v) to obtain a basis for further admission" [6]. The most common *actual* purpose for identifying causes of death is probably only "to satisfy the authorities" and make it possible to bury the deceased, i.e., purpose (iv). The medico-legal purpose may, of course, also be a more thorough forensic one. Sometimes, the physician may also have an interest to learn of the development of the disease process for the handling of similar cases in the future, purpose (ii). The investigation may also be a part of a research study, purpose (iii). The purposes (i) and (v), which concern the future handling of the particular patient, are obviously irrelevant to the identification of causes of death.

[17] Today called "Statistics Sweden".

## BIBLIOGRAPHY


1. Alderson, M.: 1981, *International Mortality Statistics*, The Macmillan Press Ltd, London and Basingstoke.
2. Buck, C.: 1975, 'Popper's Philosophy for Epidemiologists', *International Journal of Epidemiology* 4, 159–168.
3. Buck, C.: 1976, 'Popper's Philosophy for Epidemiologists' [Letter to the Editor], *International Journal of Epidemiology* 5, 97–98.
4. Creese, A.: 1975, 'Popper's Philosophy for Epidemiologists' [Letter to the Editor], *International Journal of Epidemiology* 4, 352–353.
5. Davies, A.M.: 1975, 'Epidemiological Reasoning. Comments on "Popper's Philosophy for Epidemiologists" by Carol Buck. Comment One', *International Journal of Epidemiology* 4, 169–171.
6. Elffors, L.: 1988, 'On Assessing the Validity of the Main Diagnosis in Patient Data Bases: The Impact of Aims for Making Diagnosis', *Theoretical Medicine* 9, 141–150.
7. Eyler, J.M.: 1979, *Victorian Social Medicine. The Ideas and Methods of William Farr*, Johns Hopkins University Press, Baltimore and London.
8. Francis, H.: 1976, 'Epidemiology and Karl Popper' [Letter to the Editor], *International Journal of Epidemiology*, 5, 307.
9. Gittelsohn, A. and Royston, P.N.: 1982, *Annotated Bibliography of Cause-of-Death Validation Studies: 1958–1980*. Vital and health statistics, data evaluation and methods research. Series 2, No. 89. DHHS Publication, No. (PHS) 82-1363, NCHS, Hyattsville, Maryland.
10. Greenland, S. (ed.): 1987, *Evolution of Epidemiologic Ideas: Annotated*

*Readings on Concepts and Methods*, Epidemiology Resources Inc., Chestnut Hill, MA.

11. Humphreys, N.A. (ed.): 1975, *Vital Statistics: A Memorial Volume of Selections from the Reports and Writings of William Farr* (Reprint of the 1885 ed.), The Scarecrow Press, Inc., Metuchen, N.J.
12. Jacobsen, M.: 1976, 'Against Popperized Epidemiology', *International Journal of Epidemiology* 5, 9–11.
13. Karhausen, L.R.: 1986, 'Re: "Popperian Refutation in Epidemiology"' [Letter to the Editor], *American Journal of Epidemiology*, 123, 199.
14. Lindahl, B.I.B.: 1984, 'The Reliability of Swedish Mortality Statistics for Rheumatoid Arthritis', *Scandinavian Journal of Rheumatology* 13, 289–296.
15. Lindahl, B.I.B.: 1984, 'On the Selection of Causes of Death: An Analysis of WHO's Rules for Selection of the Underlying Cause of Death', in L. Nordenfelt and B.I.B. Lindahl (eds.), *Health, Disease, and Causal Explanations in Medicine*, D. Reidel Publishing Company, Dordrecht, Netherlands, pp. 137–152.
16. Lindahl, B.I.B.: 1985, 'In What Sense is Rheumatoid Arthritis the Principal Cause of Death? A Study of the National Statistics Office's Way of Reasoning Based on 1224 Death Certificates', *Journal of Chronic Diseases* 38, 963–972.
17. Lindahl, B.I.B.: 1988, 'On Weighting Causes of Death. An Analysis of Purposes and Criteria of Selection', in A. Brändström and L.-G.Tedebrand (eds.), *Society, Health and Population During the Demographic Transition*, Almqvist and Wiksell International, Stockholm, pp. 131–156.
18. Maclure, M.: 1985, 'Popperian Refutation in Epidemiology', *American Journal of Epidemiology* 121, 343–350.
19. Maclure, M.: 1986, 'The Author Replies' [Letter to the Editor], *American Journal of Epidemiology* 123, 199.
20. Maclure, M.: 1986, 'The Author Replies' [Letter to the Editor], *American Journal of Epidemiology* 123, 1120–1121.
21. Nordenfelt, L.: 1983, *Causes of Death: A Philosophical Essay*, Swedish Council for Planning and Coordination of Research, Report 83:2, Stockholm.
22. Nyström, E.: 1988, 'The Development of Cause-of-Death Classification in Eighteenth Century Sweden. A Survey of Problems, Sources and Possibilities', in A. Brändström and L.-G. Tedebrand (eds.), *Society, Health and Population During the Demographic Transition*, Almqvist and Wiksell International, Stockholm, pp. 109–129.
23. Popper, K.R.: 1978, *Conjectures and Refutations. The Growth of Scientific Knowledge* (4th ed.), Routledge and Kegan Paul, London and Henley.
24. Popper, K.R.: 1979, *Objective Knowledge. An Evolutionary Approach* (Revised edition), The Clarendon Press, Oxford.
25. Preston, S.H., Keyfitz, N. and Schoen, R.: 1972, *Causes of Death. Life Tables for National Populations*, Seminar Press, New York and London.
26. Rapoport, A.: 1958, 'Various Meanings of "Theory"', *The American Political Science Review* 52, 972–988.
27. Sandblad, H.: 1979, *Världens nordligaste läkare. Medicinalväsendets första insteg i Nordskandinavien 1750–1810*, [with a summary in English, 'The Northernmost Doctors in the World. The Beginnings of Public Medical Care in

Northern Scandinavia 1750–1810'], Almqvist and Wiksell International, Stockholm.
28. Smith, A.: 1975, 'Epidemiological Reasoning. Comments on "Popper's Philosophy for Epidemiologists" by Carol Buck. Comment Two', *International Journal of Epidemiology* 4, 171–172.
29. Suppe, F. (ed.): 1977, *The Structure of Scientific Theories* (2nd ed.), University of Illinois Press, Urbana, Chicago, London.
30. Susser, M.: 1987, *Epidemiology, Health, & Society. Selected Papers*, Oxford University Press, New York, Oxford.
31. Weed, D.L.: 1985, 'An Epidemiological Application of Popper's Method', *Journal of Epidemiology and Community Health*, 39, 277–285.
32. Weed, D.L.: 1986, 'On the Logic of Causal Inference', *American Journal of Epidemiology* 123, 965–979.
33. Weed, D.L. and Trock, B.J.: 1986, 'Criticism and the Growth of Epidemiologic Knowledge. (Re: "Popperian Refutation in Epidemiology")' [Letter to the Editor], *American Journal of Epidemiology* 123, 1119–1120.
34. WHO: 1948, *Manual of the International Statistical Classification of Diseases, Injuries, and Causes of Death*, Vol. 1, 6th revision, WHO, Geneva.
35. WHO: 1952, *Medical Certification of Cause of Death. Instructions for Physicians on Use of International Form of Medical Certificate of Cause of Death,* WHO, Geneva.
36. WHO: 1977, *Manual of the International Statistical Classification of Diseases, Injuries, and Causes of Death*, Vol. 1, 9th revision, WHO, Geneva.

LENNART NORDENFELT

# COMMENTS ON WULFF'S, THUNG'S AND LINDAHL'S ESSAYS ON THE GROWTH OF MEDICAL KNOWLEDGE

## INTRODUCTION

Henrik Wulff's and P.J. Thung's highly interesting papers are alike in that they assume a quite general perspective toward medical research and medical knowledge. They also partly argue in the same direction, make similar observations, and draw similar conclusions. They share the view that medicine cannot be just a biological science. Medical research, they say, frequently focuses on problems which are not purely scientific in nature. Therefore, it requires a great deal of non-scientific knowledge.

Wulff uses his insights mainly as an imperative to medical scientists. Since traditional medical research has almost totally been concentrated on biological pathogenesis, it should now, according to Wulff, be partly redirected into, for instance, environmental and life-style studies. Thung's project, on the other hand, is basically one of philosophy of science. His main theme concerns criteria of progress in science in general and medicine in particular. He realizes that if medical knowledge is to encompass, for instance, psychological, sociological and ethical knowledge, then the resulting science of medicine must have new criteria of progress. He claims that there is a need for a redefinition of medical problems and medicine itself. On the *individual* level medicine must encompass the experiences of patients and physicians as well as questions concerning personal responsibility. On the level of the *population* medicine must tackle problems of social equality and effectiveness. A science of such a variety cannot be judged solely according to criteria derived from natural science.

Ingemar Lindahl's contribution to this theme is very different from Wulff's and Thung's. Instead of discussing progress in medicine in general Lindahl focuses on a quite specific issue, i.e., the assessment of the scientific value of current cause-of-death theories. He suggests a number of criteria for such an assessment and analyzes the development of cause-of-death research according to these criteria.



*H.A.M.J. ten Have et al. (eds.), The Growth of Medical Knowledge, 121–129.*

## ON HENRIK WULFF'S "FUNCTION AND VALUE OF MEDICAL KNOWLEDGE IN MODERN DISEASES"

I find Wulff's essay most illuminating and informative. I have nothing to add to the first part, where Wulff deals with topics of epidemiology and the present state of medical research. I support his demand for a partial redirection of medical research and I share his view that *preventive medicine* must gain a more prominent position than it now has. I also agree with Wulff in the latter part of his discussion where he emphasizes the need for cultural knowledge of human beings and their different life-situations in order to achieve the ultimate goal of medicine, i.e., a state of health for the population.

I am not so sure, however, that all the new research which is called for by Wulff and others is a matter for medicine. There are other disciplines, I wish to argue, which are working for *the target of health*, and which could be better suited for pursuing some of the required cultural and humanistic research. I shall indicate more clearly what I have in mind in my comments on Thung's paper.

My specific comments on Wulff's paper will be confined to his sections on hermeneutic and ethical knowledge and concern mainly conceptual matters.

My first question concerns Wulff's treatment of the very notion of hermeneutic knowledge. What does such knowledge actually consist of on his view? One thing is entirely clear. Hermeneutic knowledge, according to Wulff, is different from scientific knowledge. But is it anything more specific? One may doubt that Wulff intends anything very precise when he discusses the attitude of ordinary clinicians towards their profession. He says:

> "... clinicians often adopt the scientific view *with humanistic constraints*. They adopt the scientific view of disease as abnormal biological function, but they also show compassion and treat their patients as fellow human beings. They feel that they know which treatment is best, but they also appreciate that they must respect their patients' autonomy and secure their informed consent" ([7], p. 82).

But what does this point show? Does Wulff mean that a scientist, who is a humanist in the minimal sense of acting according to an ethical code, necessarily uses hermeneutic knowledge?

Later on in the same section Wulff makes some important observations about the concepts of health and disease. He attacks the idea that disease is simply the same as abnormal biological function. He says, in a polemic with Christopher Boorse, that the normality of the function of a human being must be viewed in relation to the life experience, hopes, values, wishes, and feelings of that individual.

With some reservations I support this conclusion [4], but what is *hermeneutic* about it? The conclusion certainly implies that extra-biological knowledge, such as psychological and cultural, must enter into the picture in making medical diagnoses, but is that knowledge necessarily hermeneutic? My general point is the following: Wulff's basic observations are too important to be covered by the very theoretically loaded and ambiguous term 'hermeneutic knowledge'. What he intends can be better expressed by the more general and neutral terms that he uses. *Humanistic* knowledge (including psychological, anthropological and philosophical knowledge) is indeed an important requisite for medicine and health care.

The term 'hermeneutics' need not be banned. But in order to be interesting and useful it should, I think, be restricted to a narrower use, for instance to referring to the field of interpretation of signs, including texts, utterances and symbolic actions. Such interpretation certainly enters into medical research and practice. But it does not exhaust the whole area that Wulff discusses.

Let me now turn to Wulff's treatment of ethics. First a point which touches the old philosophical problem about the relation between facts and norms.

In his very interesting paragraphs about different concepts of health and disease Wulff makes the very strong claim that "*paternalism is the logical consequence of the biological concept of disease*" ([7], p. 83). The rejection of this concept, he goes on to say, necessitates the participation of the patient in the decision process.

This cannot be entirely correct and it also goes against certain other passages in Wulff's paper (see our previous quotation from Wulff). Embracing a biological concept of disease is certainly compatible with respect for the patient's will. Even if the doctor claims to have the full truth about the patient's illness, he may respect that the patient does not want to be treated for his illness. This logical observation, however, should not hide Wulff's important insight on the matter. A purely scientific view of disease makes a paternalistic attitude more understandable and more reasonable. A humanistic view of health and disease – according to which a diagnosis logically requires knowledge of the patient's goals and aspirations – necessitates some communication with the patient. This in its turn makes a non-paternalistic attitude more natural. But it does not with logical necessity exclude paternalism.

My next question-mark concerns Wulff's view on the status of medical ethics. Wulff claims that medical ethics is an important discipline, but "*it is not an international one*, as, for instance, physiology or orthopaedics" ([7], p. 85, my italics). I understand the idea behind this statement. Wulff notices the cultural relativity of written and unwritten ethical codes in medicine. What is to be considered to be proper ethical conduct may partly diverge between, e.g., North American, Dutch, and Danish hospitals due to the influence of different ethical principles.

However, to use this fact in concluding that ethics is *not* an international discipline seems to me highly misleading. In reply I wish to make the following two observations.

(a) By using the same kind of argument we can prove that none of the arts or social sciences is international. They all, completely or partly, study facts of culture, which may vary from society to society. But even parts of pathology and medicine may be affected by this reasoning. Some diseases are dependent on certain constellations of background factors, a few of which may be cultural. Such constellations may be present only in certain societies. Hence, the diseases in question are essentially culture-dependent.

(b) Medical ethics is *not* exhausted by the study of current ethical codes in different societies. Ethics as a discipline contains much more. Let me just mention the following areas: the study of ethical language, the analysis of ethical arguments, the construction of consistent ethical systems, and most fundamentally, the analysis of the nature of ethics. In all these areas there is a long tradition of a truly *international* discussion.

Let me conclude by making a short comment on Wulff's view on medical philosophy. This discipline, he says, is a medical discipline, and as such it serves the same end as other medical disciplines, i.e., the elimination of diseases and the promotion of health. I do not wish to deny that this is one of the purposes of medical philosophy. There is, however, a further end. Medical philosophy, like philosophy in general, serves to *increase our understanding of the world* and, in particular, of our own intellectual enterprises in the world.

## ON P.J. THUNG'S "GROWTH OF MEDICAL KNOWLEDGE"

Thung raises a number of highly interesting issues in general philosophy of science as well as in the philosophy of medicine. He discusses the concept of progress in science and uses this discussion as a platform for a particular analysis of progress in medical science. Thung partly agrees with Wulff that medicine is not *just* a natural science. It is something more. Thung suggests that parts of psychology and sociology belong to medicine. As a result, progress in medicine cannot be judged only according to scientific criteria, whatever they are, but must be judged according to a more complex set of criteria.

I shall here only deal with what Thung says about medicine. I have general sympathy for his line of argument but I think that his analysis could be more forceful and perhaps clarified with the help of some distinctions. Let me make an attempt in this direction by "unpacking" one segment in Thung's reasoning.

"The simplest criterion for scientific progress was that science works, that it increases our grip on nature. In order to establish continued progress of medical knowledge, we should observe that medicine becomes ever more effective.... For this claim I could of course invoke such evidence as transplantation surgery, new medicines, or *in vitro* fertilization. With equal ease, however, I could be refuted by contrasting evidence: the lung cancer epidemic, rampant resistance of micro-organisms against antibiotics, high rates of infant mortality in some countries, in others the increasing frequency of intractable diseases like Alzheimer's disease, multiple sclerosis and many more" ([5], p. 94).

In this discussion Thung presupposes a quite extreme interpretation of the locution "increasing our grip on nature", in fact amounting to: improving the health of the *population*. The science of medicine, according to this reading, is progressing only if we are actually succeeding in raising the state of health of the population. This is certainly not the only possible interpretation. Consider the following alternatives:

a. *Increasing our intellectual grip on nature*. Here we mean that our understanding of nature increases. We have construed hypotheses and theories; we have put them to test and found (or decided) that our observations corroborate these hypotheses or theories.

b. *Increasing our technological ability to manipulate nature*. On this interpretation we mean that science works by giving us an effective technology. Assume that certain theories in solid state physics can be used for the technology of bridge-construction. Science then works by *enabling* us to manipulate nature in a way that we wish.

c. *Increasing our effective manipulation of nature*. Ability does not necessarily lead to action. *A fortiori* it need not lead to mass-action. A technology need not be spread and put into mass-fabrication. We may have the knowledge of an advanced and efficient bridge-building technology without being able to market it and thereby not being able to benefit from it. It is only when this technology has been diffused all over the world that we can say that a particular part of science has a grip on nature in the full-blown sense of the word.

We can now see how the triad (physics – technological science – technological application and diffusion), has a parallel in the triad (basic medical science – clinical science – health care and health promotion). There can be progress in basic medical science in the sense of a biological or psychological understanding of man, without there being a new therapeutic measure invented. There can be progress in clinical science, in the sense that a new efficient medical technology has been created, without this technology actually coming into use or being put into mass-fabrication. Great achievements can be made on the first two levels, and to some extent also on the third, without there being any noticeable improvements in the population's state of health. Thus it would be misleading to say that there is no progress in basic or clinical cancer research, just because there is a high prevalence of lung cancer. We may know perfectly well how to treat the lung cancer and even how to prevent it. Still our knowledge may have little impact because *the resources for health care are small or because the general political situation prevents effective measures against the disease.*

It is certainly true that our ultimate aim is to reduce illness and promote health in the population. For this we require a very complex stock of knowledge, much of which is not traditionally medical. We need much non-medical knowledge to answer questions such as the following: How do we organize health care most efficiently? How do we make doctors aware of new insights in clinical medicine? How do we make the general public aware of health risks? How do we change our environment to reduce the causes of disease and injury? All these are questions of *health-promotion* and *health-education.* These highly important areas have nowadays become the topic of disciplines in their own right. These, the disciplines of health-promotion and health-education, receive their intellectual input from several existing disciplines, mainly from the social sciences and the humanities.

A crucial question to be put to Thung now is: Does Thung mean that health-promotion and health-education as disciplines are or should be parts of any particular medical sciences? If they are, then the criteria of progress in medical science will be very complex indeed. If they are not, and instead continue to develop as separate disciplines cooperating with medicine, then medicine need not have all this complexity added to it. Health-promotion and health-education as social sciences or humanistic disciplines will definitely have their own mostly non-scientific criteria of progress.

## ON B. INGEMAR B. LINDAHL'S "THE DEVELOPMENT OF POPULATION RESEARCH ON CAUSES OF DEATH: GROWTH OF KNOWLEDGE OR ACCUMULATION OF DATA?"

In the course of his discussion, Lindahl makes a number of insightful remarks on the present state of the registration and statistics of causes of death. His main concern, however, is the epistemological status of cause-of-death-theories as a criterion for progress in population research. He asks: Do they serve any clear purpose? In that case, what is their purpose? Lindahl makes the claim that the whole process from the collection of data to the construction of theories must be guided by a well-defined *purpose* or a combination of well-defined purposes. He distinguishes between the following five purposes that the cause-of-death-theories may serve:

(i) to describe what caused death;
(ii) to explain why death occurred;
(iii) to identify causes for the prevention of untimely death;
(iv) to identify risk factors for the prediction of untimely death; and
(v) to adjudge personal responsibility for death ([2], p. 108).

I find these distinctions valuable. They call, however, for a *further* analysis not least in order to detect important interconnections between the purposes. It seems to me that Lindahl overemphasizes the distinctions and does not develop the interdependencies.

Let me illustrate by discussing WHO's definition of the notion of "an underlying cause of death". By such a cause of death is meant either (a) the disease or injury which *initiated* the train of events leading directly to death, or (b) the *circumstances* of the accident or violence which produced the fatal injury. WHO's purpose of proposing this definition has been expressed in the following way:

> "From the standpoint of prevention of deaths, it is important to cut the chain of events or institute the cure at some point. The most important public health objective is to prevent the precipitating cause from operating" ([6], pp. 699–700).

Lindahl points out that the WHO does not serve this purpose well. Some of their rules "for selection of the underlying cause give preference to more serious conditions over less serious and thus serve a purpose of prediction rather than prevention" ([2], p. 110). Had they really focused on the preventive purpose, says Lindahl, they should have concentrated on, e.g., irreplaceable or manipulable causes [1].

In order to make my first point, I shall propose an interpretation of WHO's quoted dictum. The clause "preventing the precipitating cause from operating" can be given at least two readings: (a) cutting the connection between the *precipitating* cause (i.e., the underlying disease or injury) and that which it *immediately* causes; and (b) preventing the precipitating cause from occurring. The latter interpretation, which is in line with the idea of primary prevention, is certainly the most reasonable. In adopting this interpretation we can make the following observation. The issue of prevention does not then rely on the manipulability of the underlying disease itself. What is to be manipulated is the external or internal cause of this disease. Consider the example of lung cancer – which is a serious disease and would most naturally be chosen as the underlying cause of death in a particular death certificate. As we know, lung cancer is at present rarely a curable disease and thus rarely an eliminable cause of death. However, the habit of smoking, which is supposed to be a major cause of lung cancer, is quite amenable to manipulation.

Secondly, I find it quite reasonable to focus on serious diseases if one's ultimate purpose is the prevention of untimely death. Consider a disease which is serious in the strong sense that it is frequent and affords a high probability of death for its bearer. (Observe that the notion of severity of a disease can be interpreted in a number of ways. For a discussion see [3], pp. 88–94). If we were to eliminate such a disease, we would make a substantial contribution to the prevention of untimely death in the population. It may be true that the disease is at present incurable, and that there are no known causes of it which are amenable to manipulation. Such a fact should, however, not prevent medical research and health care from directing all their efforts to eliminate this disease. The severity of the disease is a strong reason for an uninterrupted *search* for a cure of it or for manipulable causes of it.

The upshot of this reasoning is that we need better *explanatory* and *predictive* theories as guides in our search for important *manipulable* causes of death. Admittedly, effective prevention in the end requires the identification of some manipulable causes of death. However, in the process of finding such causes, we must be guided by adequate explanatory and predictive theories. The different purposes of cause-of-death-theories that Lindahl registers in his essay are therefore not as independent as his presentation might at first suggest.

*Linköping University,*
*Linköping, Sweden*

## BIBLIOGRAPHY

1. Lindahl, B.I.B.: 1988, 'On Weighting Causes of Death. An Analysis of Purposes and Criteria of Selection', in A. Brändström and L.-G. Tedebrand (eds.), *Society, Health and Population During the Demographic Transition*, Almqvist & Wiksell International, Stockholm, pp. 131–156.
2. Lindahl, B.I.B.: 1990, 'The Development of Population Research on Causes of Death: Growth of Knowledge or Accumulation of Data?', in this volume, pp. 103–119.
3. Nordenfelt, L.: 1983, *Causes of Death: A Philosophical Essay*, Swedish Council for Planning and Coordination of Research, Report 83:2, Stockholm.
4. Nordenfelt, L.: 1987, *On the Nature of Health*, D. Reidel Publishing Co., Dordrecht, Netherlands.
5. Thung, P.J.: 1990, 'The Growth of Medical Knowledge', in this volume, pp. 87–101.
6. WHO:1977, *Manual of the International Statistical Classification of Diseases, Injuries, and Causes of Death*, Vol. 1, 9th revision, WHO, Geneva.
7. Wulff, H.: 1990, 'Function and Value of Medical Knowledge in Modern Diseases', in this volume, pp. 75–86.

## SECTION III

# IMAGE OF MAN AND THE GROWTH OF MEDICAL KNOWLEDGE

GERLOF VERWEY

# MEDICINE, ANTHROPOLOGY, AND THE HUMAN BODY

## INTRODUCTION

The Spanish historian of medicine P. Laín Entralgo wrote: "Medicine has been at all times, in one way or another, 'psychosomatic', and so it always had to be; this in contrast with pathology" ([38], p. 15). Medical practice has indeed been characterized through the centuries (and still is) by an orientation towards man as a "psycho-somatic" totality, whether this orientation was made explicit or not. Medical theory (pathology), on the other hand, has been moving away from this practical-medical point of view throughout history, to the extent that it has expressly confined itself to the somatological perspective, especially after it modelled itself both conceptually and methodically on the natural sciences.

The practical-medical point of view, that is called "psycho-somatic" (by Laín Entralgo), is essentially the same as the so-called anthropological point of view of nineteenth- and twentieth-century anthropological or anthropologically oriented medicine, and, moreover, also the point of view of common sense, of pre-scientific experience, in the sense that on this level of experience the "psycho-somatic" unity is self-evident.

Nowadays we recognize the contrast between the two orientations – the anthropological and the strictly somatological-scientific – principally as the problem of the relation between the *practice* and *theory* of medicine, of practical medicine and laboratory medicine, or put yet another way, as the tension between two conflicting tendencies or motives in medicine past and present, namely the motive of somatological reduction (reduction of the human being to its somatological aspect) and the "psychosomatic" (or anthropological) motive (rehabilitation of the point of view of the "psychosomatic" dual unity, or of the anthropological totality).

It is precisely this conflict that defines the systematic philosophical perspective from which I will discuss a chapter in the history of present-day psychosomatic medicine, i.e., that branch of medicine which is, as it were, constitutionally doomed to dwell in the area of conflict between the above-mentioned motives, and which is thereby constantly in danger of dissolving itself – as *psycho*somatic medicine, by conforming to the ideal of somatological reduction; as *scientific* psychosomatics, by the claim (not yet realized) of



*H.A.M.J. ten Have et al. (eds.), The Growth of Medical Knowledge, 133–162.*

an epistemology of its own (von Weizsäcker), which, fully recognizing the two-sidedness of "psycho-somatic" reality, transcends and synthesizes the alternatives of scientific explanation and hermeneutic understanding.

Both tendencies may be seen in twentieth-century psychosomatic medicine, as I will illustrate with reference to the vicissitudes of the so-called Heidelberg school of psychosomatics.

My aim is to show that this Heidelberg psychosomatics, and especially that of its founder, V. von Weizsäcker, is characterized by a *structural instability*. This instability becomes understandable once we see how much this psychosomatics – committed to the presuppositions of a double-aspect theory à la Fechner – is still under the influence of Cartesian thought (like all attempts in philosophy and science to solve the "psychosomatic paradox" which was explicated for the first time by Descartes). I will argue in this connection that nineteenth-and twentieth-century double-aspect-theoretical thought is not so much a solution as a rewriting of the psychosomatic paradox, a *transposition* of it on another level of description, prompted by the conviction that "certain forms of perplexity ... seem to embody more insight than any of the supposed solutions to those problems" ([44], p. 4). And my conclusion will be that *the psychosomatic paradox is still with us* and that we have no reason to assume that the existence of a scientific psychosomatics will ever be *not* threatened, caught as it is between the Scylla of a reductionist adulteration of our common sense intuitions regarding the way we experience ourselves and others, prereflexively, as a psychosomatic unity, and the Charybdis of an amazement at the riddle of our psychosomatic existence which, at best, may be phrased as the – philosophical – riddle it is.

## THE PROBLEM OF THE BODY – THE RELATION TO THE MIND-BODY PROBLEM

It is a remarkable fact that in the long history of Western-European philosophy the soul, the mind, consciousness, have always received ample philosophical attention, while philosophical interest in the body has been marginal. Even now, something like a philosophy of the body only seems to be talked of as a remote ideal, a task for philosophy present and future of which the real significance is as yet hardly being grasped. There is no lack of attempts in that direction, of beginnings of a philosophy of the (human) body – in the past and the present – but they remain sketchy, apart from a few exceptions (on which I will comment presently). The body seems to be of interest in philosophy only as the dialectical opposite of the spirit (Schelling) or of reason [5], as that which is defined by mind, or by form (Aristotle), or, as in Descartes, as that which is, ontologically, radically different from the mind. Evidently, the definition of the body mainly comes into play where the definition of the mind-body relationship is at issue.

More detailed attention to a philosophy of the body is to be found in the work of the German philosopher Hermann Schmitz [50]. In the second volume of his extensive *System der Philosophie, Der Leib*, Schmitz develops an interesting perspective on the history of philosophy under the significant heading "On the history of the concealment and the discovery of the body". The thesis he unfolds may be summarized as follows: early-Greek man (in Homer, Sappho) enjoyed in his self-conception an unconcern which enabled him not to (re)interpret feelings as subjective states of the soul, but to experience them, without distortion, as "dark, agitating powers that grip and shake man in his bodily nature" ([50], p. 365). Since then there has been a development, culminating in Plato, which, in the interest of the emancipation of the personal I from the dictates of involuntary impulses (emotions, affects), in the interest of the discovery of the soul and of the introjection of the emotions into the soul, led to psychosomatic dualism and to a denial of bodiliness ([50], p. 365). What Bruno Snell described in his famous book *Die Entdeckung des Geistes* touches one side of a development, the reverse of which is the denial of the bodily. In short: "The discovery of the mind is the denial of the body" ([52], p. 366).

Without detracting from the fascinating history of philosophy perspective Schmitz develops, which leads us from antiquity, the Stoic tonos doctrine (and its reactualization in Kant and the romantic mysticism of nature), St. Paul, early Christianity, Kabbala, J. Boehme, Oetinger, Kant, Schopenhauer, Maine de Biran to twentieth-century philosophers such as Scheler, Plessner, Sartre and many lesser-known authors, whose work has something to offer in the way of a beginning of a philosophy of the body, I want to posit emphatically that without a "discovery of the mind" there would have been no problem of the mind-body relationship, and that means: no problem of the body. Something like a notion of the body, *soma*, had been around for a long time, of course (since early-Greek philosophy), in the distinction between living and lifeless, or (in Aristotle) between natural and artificial bodies[1]. But *the body as a problem*, I repeat, could only become of real importance after *the relationship of body and mind* had become a philosophical problem of central concern, that is to say after Descartes' redefinition of the psychophysical relationship in terms of his doctrine of the two substances. For the notion of the body implied in this redefinition – the body as a machine – was, judged by our everyday, pre-scientific experience, counterintuitive. What I have just called "the problem of the body" is precisely the problem that arose with the discrepancy between the philosophical definition of the body (the body as a machine) and of the relationship of this body machine and the bodyless mind, and the evidence of our self-experience which convinces us of the – paradoxical – psychosomatic dual unity that we are, and thereby suggests a notion of an animated bodiliness for which Cartesian philosophy no longer had any place.

It is in this problem situation, created by Descartes, that the source of what I would call "the psychosomatic motive" lies, i.e., the striving – in philosophy and medicine – for restoration, for rehabilitation, of the "natural" psychosomatic unity that was lost with Descartes. Here also lies the starting point of all later endeavours that are – in answer to the motive of somatological reduction – aimed at restoration of the "psycho-somatic (or anthropological) point of view" that I consider, following P. Laín Entralgo, to be constitutive for medical practice.

## THE PARADOX OF THE PSYCHOSOMATIC, THE PSYCHOSOMATIC MOTIVE AND THE GENESIS OF SCIENTIFIC PSYCHOSOMATICS

> It seems to me that the human mind is incapable of distinctly conceiving both the distinction between body and soul and their union, at one and the same time; for that requires our conceiving them as two things, which is self-contradictory ([62], p. 52).
>
> *René Descartes*

The recollection of "pre-psychosomatic-dualistic", i.e., pre-Platonic, early-Greek thought to which H. Schmitz invites us is useful and enlightening, but it should not allow us to shut our eyes to the fact that developments in the history of the dawning of human self-awareness cannot be undone by philosophical decree. Once we have come this long, historical way there is no return to the stage of pre-reflexive immediateness. Likewise the paradox of the psychosomatic, once made explicit, as in the passage from Descartes' letter of 1643, quoted above, cannot be effaced, nor can it be denied that even at present we stand – *nolens volens* – in the area of influence of Descartes, despite all attempts to *overcome* the riddle posed by him; that the psychosomatic paradox is still very much with us, and that the motive for solving the riddle that comes with it is still actual; the paradox seems to belong to the structure of our own, human nature, denied at most, but never resolved.

Nowadays we speak of psychosomatic affections or phenomena, of psychosomatic medicine and of psychosomatics, as if they were the most ordinary things in the world. And in a certain sense, namely to the extent that these terms refer to something like "psychosomatic reality", that is what they are: to everyday, pre-scientific experience it is almost indisputable that some phenomena of life or illness belong to the sphere of the psychosomatic, even though these technical terms were (and are) not used for them. Pre-scientific experience, common sense, favors so to speak the assumption of an "ontology of the psychosomatic". Indeed, before any attempt to unravel the connection between the mental and the somatic, we already "know" that we are, in our lived, concrete reality, a *unity* of both, of *psyche* and *soma.* This knowledge is immediately given. And our perception of others and our association with them also presuppose this unity of body and soul as something self-evident. It is the evidence of this pre-scientific experience of ourselves and others that underlies the "psychosomatic point of view" mentioned above, a point of view that may not only be regarded as constitutive for medical practice [38], but that is also, up to the present day, the starting point for every philosophical position that, in opposition to physicalistic or hermeneutic-philosophical reduction, clings to the riddle of our human, psychosomatic dual unity.

Undoubtedly common sense has its own historicity, but as long as one wants to refrain from the extreme of a thorough relativism in this, it seems defensible to distinguish, in the history of Western-European thought from its origin in Greek antiquity, movements, trends, or simply authors, according to their diverging from or converging on the common sense point of view, or the "psychosomatic point of view" contained in it.

Thus it seems plausible that the episodic dominance of the religious-moral perspective of the mystic redemption doctrines of Orphicism and of the mysteries of Eleusis in Greece (the eighth till the fifth century BC) put the common sense position under pressure, and favored, temporarily, a definitely dualistic anthropology which also left traces in the early work of Plato (427–347 BC). With Aristotle (384–322 BC) however, the pendulum swings back in the direction of the common-sense orientation: in reaction to the dualistic aspects of Plato's philosophy (dualism of body and soul) and the doctrine of the three parts of the soul that went with it, he designed, as part of his general metaphysics, a doctrine of the relationship of body and soul in which *the authorization of the common-sense point of view by philosophical means* reached a summit for a long time to come. Aristotle posited the identity of body and soul in the sense of hylomorphism[2]: every concrete individual entity is conceived as the unity of the two *principles* of matter (*hyle*) and form (*morphe*, *eidos*), and to the extent that we are talking about "natural" things like living beings (as opposed to artefacts), this means unity of body and soul. In fact, this hylomorphistic philosophy of nature represents a "psychosomatic" position in the sense that the "psychosomatic", so to speak, pertains not just to a section of (natural) reality, but to the structure of reality in its entirety: the existence of *all* living beings, including man, manifests this "psychosomatic" structure.

As I have already pointed out, Descartes' philosophy constitutes a turning point. Not only because it forms the origin and starting point of a long history of fruitless attempts to solve the mind-body problem, but especially because, by redefining this mind-body relationship as a relationship of a bodyless mind and a body-machine, in opposition to Aristotelian-scholastic thought, it drove the whole ensuing history of the mind-body problem in a counterintuitive, anti-common-sense direction; a way of thinking, if not Platonic, then surely "Platonistic", which has shaped, up to the present day, the conception of the mind-body problem in philosophy and science, particularly in the Anglo-Saxon world – notwithstanding an unmistakeable zeal in some present-day trans-atlantic physicalists (D.C. Dennett, P. & P. Churchland) who would have us believe that the mind-body problem is a pseudo-problem, and also in spite of all attempts, particularly in twentieth-century European philosophy (in phenomenology, philosophical anthropology, hermeneutic philosophy) to get rid of this "cancer growth" of modern philosophy and psychology[3].

The psychosomatic motive was to become active again in the second half of the eighteenth century – not so much (at first) in philosophy, as in medicine and psychiatry. I refer in particular to the German (medical-)anthropological tradition that culminated between 1770 and 1820. Here – certainly not by accident in medical quarters – a rehabilitation of the psychosomatic point of view takes place in reaction to Cartesianism and its consequences. So clearly does the problem of psychosomatic unity form the thematic heart of this anthropological tradition, that the defense of the anthropological outlook is all but equivalent to a plea for the acknowledgement of the psychosomatic unity of human beings[4].

Anti-Cartesianism – in German medicine, psychiatry *and* philosophy around 1800 – became the vehicle for a rehabilitation of the viewpoint of psychosomatic *unity*. It was not only the above-mentioned anthropologically-oriented medicine and psychiatry that became important here[5], but also and more particularly the so-called "romantic" philosophy of nature (Schelling,

Novalis), and the "speculative medicine of Romanticism"[6] inspired by it. It is notably in this school of romantic philosophy of nature, and more particularly in romantic mysticism of nature, that H. Schmitz discovered important origins of the "concealed" philosophy of the body I referred to earlier.

From the 1840s onwards German medicine and physiology disengaged themselves from their ties to a romantic philosophy of nature and developed into disciplines modelled on natural science. This conscious distancing from the past was carried out as a methodical reorientation: instead of systems of knowledge of nature and medicine construed on the basis of models of polarity and analogy, stress was now being laid on empirical facts, i.e., on observation and experiment. No longer was nature (i.e., organic nature) seen as Nature, which could be fathomed as the dialectical opposite or as externalization of the Spirit, by mystical turning inwards, poetic vision, or philosophical speculative reflection. There was no longer any question of the intimate belonging, the ontological oneness, of the subjectivity of the investigator and the investigated in the relationship of microcosm and macrocosm[7]. Nature as it is at issue after this reorientation is nature as we talk about it since Descartes, that is to say, nature as the essential opposite of the investigating subjectivity of the scientist. Nature is no longer the meaningful whole that invites interpretation of its hidden meaning, its sense (e.g., as externalization of the Spirit, as all-pervading Life, etc.); it is essentially *meaningless* factuality.

It does not seem strange that after the eclipse of romantic thought no psychosomatic questions were raised for a long time, once one realizes that the rise and propagation of the scientific orientation in the life sciences and in medicine since the middle of the century implied a shift in philosophical presuppositions (causal-mechanistic presuppositions, combined with materialism or naturalism). This shift led to a devaluation of the "psychosomatic-anthropological point of view" (in the sense of Laín Entralgo) of medical practice and discouraged a continuation of "romantic" attempts at a specifically hermeneutic approach to the psychosomatic problem (the relationship of psyche and soma conceived as a relationship of expression). Thus it did not leave much room for the development of specifically psychosomatic questions. Besides, given the initial predominance of materialistic, epiphenomenalistic, or parallelistic positions, the idea of *mental causation* – an indispensable element in a psychosomatics that models itself on natural science – was long thought suspect. The notion had to be made scientifically respectable before it could be pressed into service in the scientific handling of psychosomatic problems[8].

Accordingly, up to the end of the century the psychosomatic did not play any prominent role as a medical-psychological problem, either in psychiatry, or in internal medicine[9]. A strictly somatological self-understanding of medicine – the epitome of the ideology (and idolatry) of natural science oriented medicine then and now – blocked the way to any form of scientific psychosomatics. This state of affairs could and would change only at the end of the century when causalistic-mechanistic thought came under attack in the whole range of the sciences of life, from different sides[10]. Specifically holistic or organicist approaches in biology, physiology, and pathology came to the fore [43]. And with that the *machine model* of the body – a heritage from Descartes – gives way to an *organicist concept* of the body. The functional pathology of G. von Bergmann, important because it paved the way for psychosomatic medicine, continues in this line of development[11].

In fact, however, we witness the first attempts to slip psychology into the citadel of medicine much earlier, namely by way of psychiatry, which until about 1870 had been predominantly neuro-or brainpsychiatry – an important step in the emancipation from somatological dogma.

However, neither the organicist re-orientation in pathology, nor the introduction of psychology in psychiatry, particularly as it took shape in the Heidelberg school of psychiatry ([28, 60]) – first as experimental psychology in the sense of Wundt, introduced by E. Kraepelin, then as *verstehende* psychology (psychology of understanding) in the sense of Dilthey, as advocated by K. Jaspers and H. W. Gruhle ([29, 20]) – were sufficient (taken individually or together) to found modern scientific psychosomatics. To that end a conceptual and methodical re-orientation of medical thought was needed, and this meant a radical change in the self-understanding of the doctors as well as of the psychologists at the time. This re-orientation became possible with *psychoanalysis*.

The model of hysterical conversion proposed by Breuer and Freud in 1895 became the starting-point for further development of specifically psychosomatic questions [11], in the American version (F. Alexander) as well as in the German, i.e., Heidelberg version, in which the work of V. von Weizsäcker (1887–1957) in particular was to play a prominent role. In the long history of alternating recognition and want of appreciation of the psychosomatic point of view, the psychosomatics of von Weizsäcker occupies a position *sui generis* which is not always recognized as such. The analysis of that position will be the subject of the second half of this article.

## BEYOND THE PSYCHOSOMATIC PARADOX?

> Certain forms of perplexity – for example, about freedom, knowledge, and the meaning of life – seem to embody more insight than any of the supposed solutions to those problems ([44], p. 4).
>
> *Thomas Nagel*

> Nagel is both courageous and clever. It takes courage to stand up for mystery, and cleverness to be taken seriously. Nagel repeatedly announces that he has no answers to the problems he raises, but prefers his mystification to the demystifying efforts of others. Oddly enough, then, Nagel would agree with me that his tactical starting point leads not just to perplexity, but to a perplexity from which he himself offers no escape ([12], p. 5).
>
> *Daniel Dennett*

Is there a beyond of the psychosomatic paradox, the enigma of our duality which is nevertheless a unity? No, but we can try to redefine what appears to be so enigmatic: as the enigma of our unity which is nevertheless a duality. That is not a real solution, but a strategy which, though never very popular, crops up time and again, up to the present day, in the history of the mind-body problem – since the nineteenth century under the name of double-aspect theory. As an answer to the problem which was posed by Descartes it remains bound to the presuppositions of Cartesian thought (subject-object scheme), but for the rest it differs clearly from all other philosophical positions in the mind-body problem that stand in the *Wirkungsgeschichte* of Cartesian thinking, because of (1) the rejection of an unqualified ontological duality and (2) the consistent refusal to give precedence to either one of the two relata in the mind-body relationship.

An analysis of von Weizsäcker's position concerning the mind-body problem shows a similar pattern and suggests that von Weizsäcker somehow remained committed to the presuppositions of nineteenth-century double-aspect theoretical thought. Only after this hypothesis has been made sufficiently plausible, will we consider the question of what light this identification throws on the further vicissitudes of this variety of psychosomatics, in the context of post-war German and Heidelberg psychosomatics.

### *Philosophical and historical context. From nineteenth-century double-aspect theory to twentieth-century double-perspective theory*

There are recent tendencies in contemporary philosophy which, motivated by simultaneous doubts about the possibility of a full-fledged physicalism and hermeneuticism, have given rise to a revival, modest to be sure, but unmistakable, of double-aspect (or double-perspective) theoretical positions[12]. They sharpen our discernment of and raise our interest in older varieties of double-aspect theoretical thought, be they manifest or hidden. What does the theory maintain?

The double-aspect theory says that "mental and bodily facts are parallel manifestations of a single underlying reality" ([2], pp. 295–296). This position has its intellectual ancestor in Spinoza: "the decision of the mind ... and determination of the body, are simultaneous in nature, or rather one and the same thing, which when considered under the attribute of *thought* and explained through the same we call a *decision*, and when considered under the attribute of extension, and deduced from the laws of motion and rest, we call a *determination*"[13]. As Vesey [61] rightly remarks, the double-aspect theory is Spinozism popularized. God, the one substance, is dropped and attributes, with their substance-philosophy connotation, are replaced by "aspects". Double-aspect theorists feel that they are not being more than marginally metaphysical ([61], p. 146). There have been a number of different double-aspect theories, all of them to some extent starting from the threefold pattern implied by our everyday concept of "aspect": (a) that which presents the aspects, (b) the aspects themselves, and (c) the person to whom the aspects are presented. In actual practice, however, double-aspect theories often happen to depart from this everyday concept, insofar as they tend to restrict its meaning to just one element of the three that make up the original, everyday concept of aspect: the aspects themselves ([61], p. 148–149). In particular, the omission from the scene of that which presents the aspects tends to reduce the double-aspect theory to what I would call not just a minimal ontology, but an epistemological theory of double perspective.

The interest in the double-aspect theory has never been very great, but the fact that it was discussed in a separate article in Baldwin's *Dictionary of Philosophy and Psychology*[14] suggests that this position enjoyed a certain fame around 1900. The theory "professes to harmonise materialism and spiritualism" ([2], p. 295) and thereby harks back to a problem situation we find in Germany in the middle of the nineteenth century. Radical changes in the social, economic, political, as well as in the scientific and philosophical sphere, contributed to making relations of spiritualism (and idealism) and materialism, of religion and science, a problem of the first order[15]. It is no coincidence that G.T. Fechner (1801–1887), whose work so clearly was in the constellation of the reconciliation of spiritualism and natural science, is commonly put forward as a typical representative of the double-aspect theory ([8], p. 130ff). The physical and the mental, according to Fechner, are, as *two modes of appearance of the same nature*, "therefore different, because it is really One and the Same which appears differently, depending on how it is taken in by different people from different points of view. Therefore too, the material brain process appears differently from the sensations and thoughts connected with it, because the same nature which underlies both together is taken in from the outside as brain process and from the inside as mental process" ([17], p. 234)[16].

However, the problem of the relationship between religion and natural science is not merely a problem dating from the middle and second half of the nineteenth century. Just after the First World War we see how, in the context of social, economic, and political upheaval and the consequent cultural and spiritual confusion and disorientation in Europe, a revival of the problem of the relationship between religion and natural science occurs in some intellectual circles in Germany. It is difficult to be blind to the extent to which a very similar problem has had a formative influence on the self-understanding of those Heidelberg doctors who initiated the Heidelberg (anthropologically oriented) psychosomatics – L. Krehl, R. Siebeck, V. von Weizsäcker. All three of them were to deliver speeches in 1919 in St. Peter's Church in Heidelberg in which the thought of a synthesis of faith and (scientific) knowledge was expressed and the idea of a "christianizing of science" was propagated ([65], vol. 1, pp. 225–226). Von Weizsäcker even went so far as to view the decline of religion, the inability to bring about a religious revival, as one or even the only cause of the catastrophe in Europe ([65], vol. 1, p. 205). As E. Luther ([40], p. 6) remarks: in this catastrophic situation it seems to von Weizsäcker (and with him, to broad circles of the German intelligentsia) that religion is the only thing which is of lasting value and deserves to be defended. Against the dangers of materialism and communism, religion is *the* only possibility of salvation, and the unity of all religions needs to be confirmed – ideas which would become most manifest in the setting up of the periodical *Kreatur* in 1926. But, generally speaking, for those Heidelberg doctors the problem of the relationship between natural science and religion was the problem of how the Christian personalistic anthropological view could be established within the practice and theory of natural science-oriented medicine[17].

It is not just the similarity in the problem situations and in the intentions to reconcile religion and science that prompt a comparison of von Weizsäcker's position with that of Fechner, nor the fact that von Weizsäcker considered

Fechner to be one of the most important influences on his own work [70], nor the fact that (in 1922) he re-edited Fechner's *Tages- und Nachtansicht*, for which he wrote an introduction ([65], vol. 1, pp. 491–501). It is rather a combination of two mutually supporting observations concerning his work that makes us suspect a commitment to a double-aspect theory way of thinking à la Fechner, which goes further than he would perhaps have liked to admit. One observation concerns the failure of his epistemological project: the development of a viable *Erkenntnismethode* [epistemology] of his own. The other observation bears on the internal structure of the epistemology to which his own epistemological position *de facto* boils down: simultaneous defense of the natural scientific and the hermeneutic perspective.

By way of introduction I should now like to turn to (the history of) Heidelberg psychosomatics in general and von Weizsäcker's position in it in particular.

### *The Heidelberg school of psychosomatics and Viktor von Weizsäcker*

I will deal very briefly with the history of the "Heidelberg school" of psychosomatics (the expression is Laín Entralgo's): it has been described, very competently, by others[18]. The Heidelberg variant of psychosomatics had its starting-point not in psychiatry, as one perhaps would expect, but in *neurology*, conceived as part of internal medicine – a novelty at the time (about 1910) in Germany ([37], p. 188). At the beginning of the Heidelberg tradition of internal medicine stands Nikolaus Friedreich (1825–1882), one of the fathers of the pathophysiological view. The pathophysiological view would reach its zenith in his "intellectual grandson" Ludolph Krehl (1861–1937), author of the famous book *Pathologische Physiologie* [35]. Viktor von Weizsäcker was to work in Krehl's department from 1908 until Krehl's retirement in 1930, first as an assistant, then while working on his doctorate, and, finally, as head of the department ([27], p. 70). Krehl is known as the man who introduced the *biographical method* in medicine. But Krehl, like his successor R. Siebeck (1883–1965) who – with Krehl and von Weizsäcker – is one of the founding fathers of anthropologically oriented medicine, is too prejudiced by the somatological self-understanding of natural scientific medicine, consistently to take the path that psychoanalysis (which would prove to be constitutive of psychosomatics) seemed to offer those interested in a reform of medicine in the anthropological sense. Although acquainted with the work of Freud and Breuer on the treatment of hysteria by hypnosis (until then treated by the Heidelberg internists in the spirit of Friedreich and his successor Erb by way of strictly *somatic* intervention or health cures and dietetic means, respectively), Krehl was never able to bring himself to integrate expressly psychoanalytic insights. The decisive impulse to bring about that conceptual and methodical re-orientation without which psychosomatic medicine would not have originated, was to come from a doctor who was not only respected in the medical world, but who was also willing to do his utmost to get psychoanalysis introduced into medicine. Such a physician was V. von Weizsäcker, whose much-quoted words on this point are clear enough: "Psychosomatic medicine has to be *depth-psychological* or it will not be anything" ([65], vol. 6, p. 455).

The breakthrough of psychoanalysis into internal medicine took place in the years 1925/1926[19]. V. von Weizsäcker played an important part in it. Of decisive importance, however, was the integration of the Heidelberg pathophysiology of the school of L. Krehl with psychoanalysis. It was von Weizsäcker's 1933 study *Körpergeschehen und Neurose* [68] which accomplished this integration, extrapolating the psychoanalytic concept of neurosis to the domain of organic diseases – Freud giving his qualified assent. Heidelberg psychosomatics had been born.

The new direction that the development of internal medicine along these lines took can be described as a sort of *subjectivization* of physiology and medicine and was accordingly carried out under the motto of "the introduction of the subject into medicine". The programmatic demand connected with that slogan in fact had two sides which, although closely connected with each other and most of the time not sharply separated by von Weizsäcker and his followers, should – at least for the sake of analysis and discussion – be clearly differentiated: an *epistemological-methodological* and a *conceptual-ontological* side.

The programme of the introduction of the subject into physiology and medicine is, in the first place, indissolubly connected with a special epistemology which von Weizsäcker considered to be essential to it ([36], pp. 90–91). An epistemology which, beyond the alternative of causal explanation and understanding [*Ein- und Ausdrucksverstehen*], indeed, as a synthesis of both ([71], p. 553), is directed towards "the comprehension [*Begreifen*] of the language of the life processes and makes their metaphorical sense possible" ([36], pp. 90–91); an epistemology of which the description, application, and experimental foundation (compare his work *Der Gestaltkreis*, [64]) occupied him during his whole life. The conceptual, ontological side of the demand of "subjectivization" becomes manifest in its ultimate consequences in what I would describe as von Weizsäcker's *"bio-logical" relationism* ("bio-logy" in the sense of a doctrine, possibly philosophy, of the living) which has its center in the category of *Umgang* (dynamic interrelation, intercourse) as the basic concept or model of relation. The *Umgangs*-relation, in the universal extension ultimately given to it by von Weizsäcker, concerns every relationship in which two "quantities" are related to each other as two poles, in a unity of interaction which produces a distinct *Gestalt. Umgang* in this sense not only occurs in the relation of perception and movement (such as is demonstrated in sensory physiological experiments [64]), but also in the relation of animal and environment, of psyche and soma, of man and nature, and finally, in dyadic relations among people (e.g., doctor and patient). "The categories of the biological are not only subjective, but also social" ([64], p. 272). The sister-concept of the category of *Umgang* concerns the unity of life conceived as an "object containing a subject" [*subjekthaltiges Objekt*]. I call it a sister-concept because both concepts refer to each other: *subjekthaltige Objekte* only exist in a *Umgangs*-relation, and every *Umgangs*-relation is one of *subjekthaltige Objekte*[20].

Viewed against this background it can be understood what, from the perspective of pathology and practical medicine, "the introduction of the subject" can mean: the "objectively" physiological or pathophysiological process assumes, so to speak, a "subjective" appearance, namely as an expression of the subjectivity of life[21]. From this time onwards, the phenomena of life and disease can – also – be conceived in terms of subject-categories. The unity of life is, after the "introduction of the subject", not merely the "objective" organism as it is studied in its functioning in natural scientific physiology, but a *subjekthaltiges Objekt*, that is, an object which is at the same time a subject.

The re-orientation of internal medicine in Heidelberg which was achieved in this way, created a stir and provoked resistance as well as criticism especially because of its *methodological implications* (particularly among doctors). The "introduction of the subject" surely meant not only a conceptual revision, introducing subject-categories into the "objective" natural science-oriented medical thought, but also the introduction of a *hermeneutic* approach to disease phenomena. The diseased organs speak a language which has to be deciphered, the disease is an expression of an *unlived life*, of repressed and unsolved problems of life. In practice this makes the doctor a psychotherapist in disguise, if not a pastor.

In that way it becomes possible to view the relation of the bodily and mental sides of man as connected in the unity of a "community of expression" [*Ausdrucksgemeinschaft*]. A state of affairs which was aptly summarized in the phrase: "Nothing in the organic sphere is without meaning, nothing in the mental is without body" ([67], p. 314). Applied to disease phenomena this meant: every somatic disease has a mental side, and *vice versa*. It is the idea of the "symbolism of expression of the language of the organs", the pet of specifically German psychosomatics at the time, but rejected by the American psychosomaticists ([63], p. 9). Disease is, from that point of view, not a meaningless event which befalls someone, but, as an expression of life problems not overcome, something for which one is personally *responsible*. That is in fact nothing short of a revival of the early nineteenth-century sin-theory of disease as we find it expressed in Windischmann and Heinroth: disease as a manifestation of an ontological deficiency of man, of his sinful nature; as something which has meaning, namely in the first place as expressive, symbolic meaning, and in the second place because, revealing what has gone wrong in the life of the patient, it simultaneously points the way to recovery. The disease, so von Weizsäcker literally says, "has the meaning to direct the patient to the meaning of his life" ([65], vol.6, p. 464), – a point of view, he adds, to which natural scientific medicine has remained completely blind[22]. This, then, in a nutshell, is the history of the Heidelberg school of psychosomatics and von Weizsäcker's role in it.

## *Towards a systematic misinterpretation of V. von Weizsäcker's psychosomatics*

### *First step: in defense of an unsympathetic reading*

Nothing in this rendering of von Weizsäcker's position seems to endorse the previously voiced hypothesis that his thought was committed to a double-aspect theory à la Fechner, perhaps even more than he would have liked to admit. This is correct: the possible meaning of that hypothesis can only be caught sight of if we change the perspective of the neutral, benevolent spectator to that of an – admittedly – unsympathetic critic. A certain measure of critical distance seems called for in the face of the uncritical acceptance and admiration one feels pervades so much of the (interpretative) von Weizsäcker literature.

It is an indisputable fact that the scientific value of the Heidelberg variant of psychosomatics connected particularly with the name of von Weizsäcker and his followers has been quite differently assessed. To insiders, sympathetic to its approach, it appeared as a revolutionary reform which released medicine from its nineteenth-century natural scientific, causal-mechanistic bias and effected its reorientation to what was to become known, after the Second World War, as von Weizsäcker's project of an *anthropological medicine*[23]. In the eyes of others (both in Heidelberg and elsewhere) this so-called "reform" appeared as a regrettable revival of an unscientific, speculative, "romantic" past which, far from advancing the cause of medicine, represented a retrograde step to the level of pseudo-science.

The severe criticism provoked by the Heidelberg variant of psychosomatics came in the first place from psychiatry ([29, 33, 63]), a psychiatry which continued the methodological self-understanding of that self-same Heidelberg psychiatry (K. Jaspers, H.W. Gruhle, K. Schneider) that had already deplored psychoanalysis as a backsliding into "romantic" psychiatry[24]. The parallel with the criticism of psychoanalysis is obvious, firstly because psychoanalysis itself was constitutive for the young psychosomatics, and secondly because in terms of intellectual history both psychoanalysis *and* Heidelberg psychosomatics à la von Weizsäcker, were actually connected, albeit in different ways, with that past of a "romantic" philosophy of nature[25].

In a scientific discussion the antecedents of the participants should not be allowed to play a part. We can leave aside the too easy identification of "romantic" with "unscientific" – a questionable product of nineteenth-century positivism – as that which it really is: a cliché which does not clarify much and has been rendered out of date by more recent history of science research. But that does not mean that the previously formulated criticism can be dismissed as misinterpretation of von Weizsäcker's intentions. That criticism was of a methodological nature[26]. Leaving aside what might have been related to von Weizsäcker's claim of an epistemology of his own and to the conceptual framework (ontology) connected with it, but recognizing the meaningfulness of the biographical-historical-(meaning) perspective concerning diseases, this methodological criticism was directed against the arbitrariness and fancifulness of the hypotheses which von Weizsäcker too often felt tempted to put forward in the application of the biographical method: "One reads with some astonishment his medical histories, is on the way to thinking anything possible, and yet in the end knows nothing", Jaspers sighs ([29], p. 208). H.J. Weitbrecht, the author of a sustained and detailed criticism of the Heidelberg psychosomatics [63], also subscribed to this methodological criticism. Standing as a spokesman for psychiatric research into the psychoses he attacked the uninhibited tendency to psychologize, displayed in the Heidelberg psychosomatics, which went so far as to interpret all forms of mental disorders – neuroses *and* psychoses – in terms of unassimilated problems of life, i.e., as the expression of life crises[27]. Such a criticism became possible, indeed imperative, not because of an inadmissable, violent misinterpretation of von Weizsäcker's intentions, but precisely because those intentions – von Weizsäcker's project of an epistemology of his own beyond the alternative of causal explanation and (hermeneutic) understanding – appeared to have no scientifically satisfactory or methodologically useful meaning apart from an unconditional commitment to that specifically Weizsäckerian "bio-logical" relationism mentioned above.

It therefore was not only understandable, but quite to the point, that the psychiatric and medical criticism focused upon that methodic aspect of Heidelberg psychosomatics in which it differed from the American psychosomatics of Alexander *et al.*: the *hermeneutic* aspect. From the standpoint of von Weizsäcker *et al.* this one-sided emphasis must undoubtledly appear as a *misinterpretation*, but taking into account the scientific claims of that psychosomatics, it was (and is), I repeat, a perfectly reasonable "misinterpretation". Once one has taken this step – and I do not hesitate to do so – the hypothesis that von Weizsäcker's thought in fact represents, *malgré lui*, a *transformation* of nineteenth-century double-aspect theory becomes plausible. In order to further strengthen the systematic "misinterpretation" of von Weizsäcker's psychosomatics I have in mind, it now is necessary to go into the question of von Weizsäcker's position on the mind-body problem.

*Second step: our unsympathetic reading continued*

> Also there, where we cannot accompany the psychological drama up to the point where it seems to turn into the somatic, perhaps continues it as its representative, also where the explanation of the somatic process no longer explains or clarifies anything, where the psychological appearance moulds everything and represents everything in a dominant way – also at these limits of psychosomatics we would not like to deny that a hidden connection exists; indeed we are somehow compelled to look further for it ([71], p. 602).
>
> The main point in the relationship of body and soul does not exist therein, that they are two things, which are there side by side, and act upon each other, but that they elucidate each other mutually...([69], p. 58).
>
> *V. von Weizsäcker*

Within certain limits the defense of the *hermeneutic* perspective in medicine makes sense – that much seems to be conceded in the psychiatric criticism referred to above. Von Weizsäcker's mistake, however, was that he did not always hold those limits in respect and that his hypotheses, accordingly, were insufficiently confirmed, not adequately supported. That means that the criticism concerned von Weizsäcker's psychosomatics insofar as (and not because) it was committed to the defense of the biographical-historical-(meaning) perspective, in short, the hermeneutic perspective. None of those critics would, if pressed, have denied that there was also another perspective – the natural scientific, causal one – which played a part in von Weizsäcker's psychosomatics. But that seemed less interesting at the time than the issue of *the scientific character of the hermeneutic approach.*

The full meaning of the causal perspective (and its relation to the hermeneutic perspective) in von Weizsäcker's thought becomes clear once one takes his views about the mind (or better, soul)-body relationship into account. It goes beyond the scope of this article to delve into this complicated subject. Fortunately, this has already been done in the Munich dissertation of H.D. Reiner [45], as far as I know the only sizeable systematic study on this theme[28]. I therefore restrict myself – by way of a shortcut – to commenting upon Reiner's main hypothesis.

Reiner attempts to show that von Weizsäcker's views about the body-soul relationship fit into the scheme of a particular *causal* theory. Though this causal theory is not explicitly defended by von Weizsäcker, it is one that, nevertheless, would be in full harmony with his ultimate intentions, so Reiner seems to suggest. To be more precise, although von Weizsäcker (read:as a medical practitioner) set great store by the hermeneutic perspective concerning the body-soul relationship (see the above-quoted passage from *Anonyma*), on closer inspection he appears to have stuck to the idea of an unrestricted validity of the causal viewpoint at all levels of the psychophysical relation, as described by him on several occasions (see the quotation above from *Der kranke Mensch*)[29].

My thesis is that Reiner's suggestion of an overall causal theory underlying von Weizsäcker's thought is to be *rejected*, and that the passages from von Weizsäcker heading this (sub)section, both quoted in connection with each other by Reiner himself ([45], p. 74), far from backing up his views, support the hypothesis that von Weizsäcker's position, when all is said and done, boils down to a double-aspect theory position which has its model in Fechner.

Von Weizsäcker's views about the body-soul relationship have been summarized by himself in a threefold scheme of what he himself preferred to call "representations" or "pictures" of the psychophysical relation ([65], vol.

9, pp. 157 and 530). The psychophysical relation can, according to this scheme, be understood in terms of (1) psychophysical causality (interactionism of psyche and soma); (2) parallelism "up to and including a together-lying unity" (body and soul in the unity of the *Leib*, i.e., the animated body; body and soul in a "community of impression and expression" [*Ein-und Ausdrucksgemeinschaft*], which means that the body is capable of expressing the experience of the soul (it is *ausdrucksfähig*), as, conversely, the soul is capable of experiencing what happens to the body (it is *eindrucksfähig*)); (3) the reciprocal representation of body and soul (this describes the clinical experience that one disease, with mental or somatic symptomatology, takes the place of another disease, with somatic or mental symptomatology).

These three forms of the psychophysical relation von Weizsäcker correlates with three *epistemological* approaches: (1) psychophysical causality corresponds with *causal explanation*; (2) the impression and expression relationship in respect of the animated body corresponds with *understanding*, and (3) the rhythmic change in reciprocal substitution corresponds with *comprehension*. "In comprehension causal explanation and understanding are included, connected, and overcome as preliminary stages" ([71], p. 553)[30]. Note that the division into three masks the fact that the basic division is a bi-partite one, namely that between the *causal* perspective and the *hermeneutic* perspective. The latter is once more divided into two sub-forms.

Reiner contends that the scheme of the three representations or pictures of the psychophysical relation can be *reduced* in its entirety (including the two hermeneutic variants just mentioned) to a psychophysical *causal* relation [45]. This causal relation, which is basic to all three of the pictures, has, according to Reiner, to be interpreted in the sense of a refined interactionism, the so-called "double-causes double-effects theory", which originated with C. Stumpf and was modified by E. Becher (1911) and A. Wenzel (1933)[31].

It is neither practical nor necessary to go into the details of Reiner's argument. Suffice it to say that the main reason why Reiner's reconstruction (in spite of relevant observations and careful documentation) does not carry conviction is the following: What von Weizsäcker does and Reiner does not see, is that there is an essential difference between external relations (in the sense of causal relations) and internal relations – relations of semantic or symbolic import[32]. The acknowledgement of such a difference, implicitly present in von Weizsäcker, cannot be reconciled with an attempt at reduction of *all* forms of psychophysical relation to the causal relation. The difficulties such an attempt at reduction must encounter are predictable: Reiner obviously does not know what to do with the *hermeneutic* variants of the forms of psychophysical relation, is disposed to read either too much causality and/or too little hermeneutics in von Weizsäcker's texts, and he consequently omits to discuss the problem of the relation of the causal *and* the hermeneutic perspectives. The conclusion to be drawn from all this (and the above-quoted passages from von Weizsäcker are an excellent illustration of it) is that the position inherent in von Weizsäcker's view of the body-soul relationship, defined in terms of the two above-distinguished methodical perspectives – the causal and the hermeneutic – *is not one of an either-or*, but of an *as well as*. In other words, what makes Reiner's study relevant to my argument, is that it (unintentionally) illustrates the point that it is impossible to give a consistent natural scientific-causal interpretation of von Weizsäcker's views without reducing the hermeneutic part of the story in an inadmissible way.

Indeed, so we may conclude our "systematic misinterpretation" of von Weizsäcker's psychosomatics, what remains after all is said and done is a psychosomatics in which *de facto* the psychosomatic relation wants to be understood neither as simply a causal relation, nor as simply a relation of expression, a symbolic or semantic relation. A psychosomatics in which both perspectives exist simultaneously, side by side, and which, though not always perhaps in the way of direct and explicit borrowing, at least on the level of its philosophical presuppositions, partakes of a Fechnerian legacy – not only epistemologically, but also ontologically. For in spite of all von Weizsäcker's disclaimers concerning possible ontological implications of his tri-partite scheme of the psychophysical relation, that psychosomatics is clearly rooted in an ontology. It is an ontology of the living, which defines the unity of living as a "subject-containing object". If one appreciates that the "subject-containing object" is essentially identical with the animated body, the *Leib*, and is on the smallest scale represented by the cell, which is conceived as "unconsciously animated", then also in point of ontology the proximity of Fechner is evident[33].

It is this ontological footing which accounts for the structural two-sidedness of von Weizsäcker's thought; which makes us assign to him a place in the history of double-aspect theoretical thought and which, in conclusion, gives the psychosomatics established by him, given its concept of the body as "subject-containing object", a position *sui generis* in the history of the changing concepts of the body which, so to speak, accompanies – underground – medical thought of the nineteenth century up to the present day. A position between the natural scientific concepts of the body of biological medicine (i.e., the body as machine, as organism, or organismic system) and the concepts of the body such as those which, originating in German philosophical anthropology at the end of the 1920s (Scheler, Plessner), finally became particularly prominent in the French, phenomenologically oriented philosophical anthropology of Sartre and Merleau-Ponty: the body as animated body, as *corps-sujet* ([22, 50], pp. 596–601, [53])[34].

## EPILOGUE

The double-aspect theory as a familiar nineteenth-century figure of thought, "while professing to harmonise materialism and spiritualism, occupies a position of somewhat unstable equilibrium between the two, and shows a tendency in different expositors to relapse into the one or the other" ([2], p. 295). That is, it shows a tendency to slide back either into psychical monism (spiritualism), or into physical monism (materialism), because either the states of consciousness or the brain processes are given the role of things that present the aspects ([61], p. 149).

Small wonder that von Weizsäcker's psychosomatics, being essentially a double-aspect theory, also shows a comparable *structural instability*. This instability tends to change it into either natural scientific positions (on a physicalistic, epiphenomenalistic, or interactionistic basis) or into a hermeneutic position (hermeneutics of nature, hermeneutics of the body). Accordingly, methodological considerations aside, the criticism leveled at von Weizsäcker's psychosomatics goes two ways: natural science oriented doctors complain about "too much subjectivity" (the hermeneutic aspect)[35], whereas a philosopher like Heidegger, spokesman for a hermeneutic philosophy, criticizes von Weizsäcker's programme of "the introduction of the subject" because the subject (conceived as a "subject-containing object") is interpreted in a natural scientific sense ("too much as an object") ([25], p. 249).

Considering V. von Weizsäcker's contribution to anthropological psychosomatics from the perspective of the history of the Heidelberg school of psychosomatics before and after him, one is compelled to say that his position is one of a *systematic "ambivalence"* or *two-sidedness*, whereas the positions of founders like Krehl and Siebeck clearly show inconsistencies stemming from a "real" ambivalence, that made them at times spokesmen for either the natural science or the hermeneutic (biographical-idiographical) point of view. Indeed, if there is one thing in which von Weizsäcker, who, incidentally, often appears to be unsystematic and doubting, has been consistent, it is in *his refusal to give absolute precedence to either the causal or the hermeneutic point of view*. Therefore, his position is actually one *between* a causal psychophysics (psychophysiology) and a hermeneutics of nature or of the body. That is why it seems only natural to relate von Weizsäcker's views to a double-aspect theory which is, although undoubtedly not identical, at least comparable to the 19th-century *Zweiseitentheorie* that found its classic expression in the work of Fechner.

The structural instability of von Weizsäcker's position was to become manifest in due course, in different ways: in the recurring *misinterpretations* of his work which tended to stress one or the other side of his work, in the, at times, massive criticism, but also in those thinkers who continued his (theoretical and practical) work, but usually took a different stance from his[36].

Broadly speaking, contemporary psychosomatics after von Weizsäcker's death (especially in Germany) proved a – modified – continuation of the psychosomatics "before the crisis", in which either psychoanalytic (A. Mitscherlich, J. Cremerius) or systems theory approaches dominate (Thure von Uexküll *et al.*)[37]. That is to say, either the hermeneutic (or psychotherapeutical), or natural science oriented approaches have determined the course of the post-Weizsäckerian development in psychosomatic theory and practice. A consistent "as well as", as was characteristic of von Weizsäcker's conceptual and methodological position, has not found a lasting following.

## PSYCHOSOMATIC MEDICINE AND THE GROWTH OF KNOWLEDGE

In 1954, E.D. Wittkower wrote: "As a field of research psychosomatic medicine has gained coherence and momentum only during the last twenty years. It is not and never can be a specialty, and the techniques and facts which it, as a movement, may accumulate are fated to be absorbed into other fields of practice till it no longer exists" ([74], p. vi). These words seem almost prophetic if one compares them with the purport of Th. von Uexküll's preface to the leading *Lehrbuch der psychosomatischen Medizin* [55][38]: psychosomatic medicine is no longer a special subject ["Spezialfach"] dealing with a limited number of so-called "psychosomatic" diseases. The subject has split up into a number of sub-disciplines. In reaction to this new situation two different answers have emerged, one going the way of furthergoing specialization (in method and subject), the other (represented by von Uexküll himself), advocating a programme of disciplinary integration[39].

The important point in this is the following: contemporary psychosomatics has in many ways outgrown von Weizsäcker's beginnings. This fact could, in a not explicated common-sense meaning of the term, be described as proof of "growth of scientific knowledge". However, one cannot be blind to the fact that this so-called "growth" or "progress" in psychosomatics in the post-Weizsäckerian area has been bought at the price of conceptual *simplification*[40] which, far from "solving" or clarifying the psychosomatic relationship, tends to reduce and to mask its problematic nature. This suggests that "growth of knowledge" in psychosomatics is only possible at the price of conceptual reduction, that is at the price of an underestimation of the force of the psychosomatic paradox.

Viewed against that background one wonders whether it could not be the case that attempts in medicine, such as presented in von Weizsäcker's psychosomatics, on the one hand, and in philosophy, as represented in versions of double-aspect theory (old and new), on the other, at least have the advantage over other conceptions of psychosomatics and philosophy – that they accept in full consciousness the psychosomatic problem as *the* problem it always was and still is: as an unsolvable mystery.

I would therefore suggest the following conclusion: though V. von Weizsäcker's programme of a psychosomatics "after the crisis" has not proved to be viable beyond the initial stage, and scientific psychosomatics – in Heidelberg – for the greater part has gone in other directions than he himself would have liked to hope or expect, the lasting value of his tireless efforts in giving the medicine of his time a new scope and vision is that it impresses upon us the unsolvable riddle which is the sphinx of our own human, psychosomatic existence. To have renewed that insight may count as growth of knowledge, if not in the domain of science, then at least in that of wisdom.

*University of Nijmegen,*
*The Netherlands*

NOTES

[1] See ([1], B1,192b13ff).
[2] leaving aside complications arising from the so-called doctrine of the *nous*.
[3] So called by the Swiss psychiatrist L. Binswanger.
[4] At the same time and in the same context (as far as I can trace, for the first time) the term "psychosomatic" (or rather: "psychic-somatic") turns up in the *Lehrbuch der Seelenstörungen* (1818) by the physician-theologian J.Chr.A. Heinroth (1773–1843);

the term "somato-psychic" appears in the work of the psychiatrist K.W.M. Jacobi (1822). It is certainly no coincidence that the occurrence of the term "psychosomatic" is (geographically and historically speaking) most notable in the German-speaking area in the period of about 1820 to about 1845: in the works of the well-known representatives of the so-called somaticist school of anthropological medicine and psychiatry, such as F. Nasse (1822, 1838), F. Groos (1828), J. B. Friedreich (1836), E. von Feuchtersleben (1838, 1845). (Dates refer to the publications cited in [41].) If Heinroth's use of the term (in 1818) was still incidental, a few years later, in Nasse, things have changed. At first (in 1820) Nasse called his system "Psycho-Physiology", and referred to it as "a doctrine for which I lack the appropriate word". Two years later, in 1822, he suggests the name "Psycho-Somatology" (with some reservations), ending up, in 1823, with "Anthropology", as "the doctrine of man". ([49], p. 20).

[5] See ([59], Ch. 1).

[6] A lucid synopsis of the views central to German Romanticism, with up to date bibliographical information, is presented in [14].

[7] G. Gusdorf ([21], p. 153ff) appropriately speaks about "anthropocosmomorphisme". See also the title of H. Schipperges [49]: "Kosmos Anthropos".

[8] See [76].

[9] I leave aside here the extra-university circuit of alternative medicine.

[10] Something not unrelated to the energetics-debate in the natural sciences in the 1890s.

[11] Culminating in [4].

[12] See Nagel [44], who in this connection refers to philosophers like Strawson, Hampshire, Davidson, O'Shaughnessy.

[13] Spinoza, *Ethics*, Note to proposition II of Book III, cited in ([62], pp. 140–141).

[14] I quote from the reprinted edition of 1960 (=revised edition of 1925). See [2].

[15] Think of the notorious materialism conflict at the beginning of the 1850s (see [59], pp. 73, 83, 175, 184, 239n283, and [9], pp. 161–188), of the storm around Darwin's *The Origin of Species* (1859) and *The Descent of Man* (1871), and also of the psychology without soul conflict since 1875 (see [59], pp. 47, 178, 183).

[16] See also ([16], Vol. I, p. 412 and Vol. II, p. 320) – cited in [26].

[17] In [58] I have analysed the religious motives underlying (among other things) the Heidelberg variant of anthropologically oriented medicine. See also [10].

[18] See particularly the Heidelberg dissertation of M. Kütemeyer [36]; see further [27] and [24], both dealing exclusively with V. von Weizsäcker.

[19] Neurology congress, Kassel, 1925; Erste ärztliche Kongress fur Psychotherapie, Baden-Baden, 1926.

[20] Cf. V. von Weizsäcker: "The monad and its encounter can be described as the *a priori* of the world" ([69], p. 61).

[21] It is quite astonishing that until now the relation of von Weizsäcker's work to nineteenth- and twentieth-century "Lebensphilosophie" has found such scant attention. Only D. Wyss (in [72]) discussed summarily the relation to the work of L. Klages. See further [32]. See also note 24.

[22] Similar views are expressed by the internist A. Jores; see for instance [30] and [31].

[23] V. von Weizsäcker differentiated between a preparatory psychosomatic medicine

*before* the crisis (i.e., psychosomatic medicine which – as in the school of S. Alexander (Chicago) – was content with being merely a supplement to natural scientific medicine) and a psychosomatic medicine *in* and *after* the crisis, which purported to achieve a wholesale revision of medicine in its entirety. The latter he called, for the sake of clarity, "anthropological medicine". See [65], vol. 7, pp. 255–282, and especially pp. 268–271.

[24] See particularly [19]; see also [60].

[25] V. von Weizsäcker never made a secret of his affinity with romantic philosophy of nature. This relation, although obvious, has as yet not been systematically investigated. The philosophical intuitions about life and bodiliness which are contained in romantic philosophy of nature and mysticism of nature, as well as such themes as the relationship of nature and spirit, life and death, health and disease, give many starting points for such an investigation. – But in its own way psychoanalysis also resumes motives ([13], pp. 199, 204, 206, 215, 514, 516, 536–537, 542, 648) and thought patterns [42] from romantic philosophy of nature. O. Marquard's thesis that psychoanalysis is a continuation of (in this case romantic) philosophy of nature with the means of science deserves serious attention in this connection.

[26] See for instance ([29], pp. 193n1, 199, 208, 567–568) and particularly [63].

[27] The controversy between psychosomatics and psychiatry crystallized around the theme "life crisis or endogenous psychosis". The two most important arguments of the psychiatrists were: (1) even the development of "normal", healthy people, is characterised – much more than agrees with our idea of a meaningful continuity of our biographical life history – by incomprehensible radical changes ["unverstehbare Umbrüche"] in the biological foundations of our existence ([63], p. 81); (2) an overwhelming majority of carefully investigated disease histories of psychotics from the psychiatric clinic shows "the psychologically speaking senseless breakthrough of the endogenous psychoses in the continuity of human existence and the *brutal interruption of meaningful connections*" ([63], p. 90).

[28] I owe this reference to Dr. S. Kasanmoentalib, who was also, thanks to her expert knowledge of the work of V. von Weizsäcker, helpful in answering a number of questions about von Weizsäcker's work that arose in the preparation of this article. Her impressive study about the scientific and philosophical background of von Weizsäcker's *Gestaltkreis* theory. [32] – deeply sympathetic to von Weizsäcker's intentions – deserves to be mentioned as a counterpart of my own "unsympathetic" reading. In fact, in so far as it helps to lay the foundation for a fair and reasoned appraisal of von Weizsäcker's work, it very happily supplements my own efforts to come to grips with this at times enigmatic author.

[29] So also where this relationship is described by him in hermeneutic terms such as *Ausdrucksgemeinschaft* [community of expression].

[30] Reiner [45] overlooks this important correlation.

[31] This refined interactionism supposes that the cortical process has effects on the mental as well as on the physical side, and that physical, neural processes have a continuous influence on the mental processes and the other way around ([45], p. 15). The partial parallelism and the partial correlation of mental and physical events which emerges in this way is nothing but a consequence of this double causes/double effects relationship. In other words, the parallelism is secondary, because it is a parallel

connection brought about by interaction, i.e., essentially an interaction-generated parallelism ([45], p. 16).

[32] See for instance ([75], p. 93) for a summary of the discussion about the distinction.

[33] "We understand the body as an unconscious and animated corporeal thing. We have good reasons to conceive all cells of the body as animated. We want to imagine every cell of the liver, every ganglion cell, every blood cell as unconscious and animated" ([65], vol. 6, pp. 402–403).

[34] Though the *terms* "subject-containing object" and "*corps-sujet*" suggest identity of meaning it should be kept in mind that the concepts of subjectivity of both relate to different philosophical frameworks: that of von Weizsäcker points to Leibnizian-Fechnerian origins, that of Merleau-Ponty to Husserlian phenomenology. Basic to the concept of *corps-sujet* is the phenomenological concept of so-called *fungierende Intentionalität* [the hidden functions of intentionality]. Indeed, from the standpoint of phenomenologically-oriented (hermeneutic) philosophy von Weizsäcker's "subject-containing object" is still too much of an object ([25], p. 249). Speaking in terms of S. Kasanmoentalib's important distinction of "organic" and "pathic" subject(ivity) in von Weizsäcker, I refer in the foregoing to von Weizsäcker's concept of "organic" subjectivity. See [32], Ch. XIII.

[35] For example ([46], p. 159) and ([47], p. 319).

[36] P. Christian continued the neurological, A. Mitscherlich the psychotherapeutical, and W. Kütemeyer (who, of the older generation, probably most approximated von Weizsäcker's ultimate intentions) the anthropological trend in Heidelberg psychosomatic thought. Also W. Bräutigam, and the younger P. Hahn, cannot properly be characterised as orthodox von Weizsäcker followers. At the moment only D. Janz and colleagues (Berlin) are, as far as I know, at least in medical practice, orientated to the model of psychosomatic medicine "after the crisis" (see [37]).

[37] See [6, 11, 15, 51, 54, 55], and [56].

[38] First edition 1970; second edition 1979; third edition 1986.

[39] For the latter see also [37].

[40] Attempts at redefining the psychosomatic problem in systems-theory terms in order to make it scientifically tractable generally suffer from this mistake. The conceptual leap from system-functional relations to semantic relations cannot be bridged by interpolation of "meaning leaps" [Bedeutungssprunge] between different levels of integration, as suggested by Th. von Uexküll and W. Wesiack ([57], pp. 68–69).

## BIBLIOGRAPHY


1. Aristotle: 1955, *Physics. A Revised Text with Introduction and Commentary by W. D. Ross*, 2nd ed., Clarendon Press, Oxford.
2. *Baldwin's Dictionary of Philosophy and Psychology*: 1960, 5th ed., (1st ed. 1901), Peter Smith, Gloucester, Mass.
3. Becher, E.: 1911, *Gehirn und Seele*, Carl Winter, Heidelberg.
4. Bergmann, G. von: 1932, *Funktionelle Pathologie. Eine klinische Sammlung von Ergebnissen und Anschauungen einer Arbeitsrichtung*, (2nd ed. 1936) Springer Verlag, Berlin.

5. Böhme, H. and Böhme, G.: 1985, *Das Andere der Vernunft. Zur Entwicklung von Rationalitätsstrukturen am Beispiel Kants*, Suhrkamp Verlag, Frankfurt a.M.
6. Bräutigam, W. and Christian, P.: 1973, *Psychosomatische Medizin*, Georg Thieme Verlag, Stuttgart.
7. Bumke, O. (ed.): 1932, *Handbuch der Geisteskrankheiten. Neunter Band: Spezieller Teil V: Die Schizophrenie*, Redigiert und mit einem Vorwort versehen von K.Wilmanns, Springer Verlag, Berlin.
8. Busse, L.: 1913, *Geist und Körper, Seele und Leib*, 2nd ed. (1st ed. 1903), Verlag von Felix Meiner, Leipzig.
9. Chadwick, O.: 1975, *The Secularization of the European Mind in the Nineteenth Century. The Gifford Lectures in the University of Edinburgh for 1973–4*, Cambridge University Press, Cambridge/London/New York/Melbourne.
10. Christian, P.: 1952, *Das Personverständnis im modernen medizinischen Denken*, J.C.B. Mohr (Siebeck), Tubingen.
11. Cremerius, J.: 1978, *Zur Theorie und Praxis der Psychosomatischen Medizin*, Suhrkamp Verlag, Frankfurt a.M.
12. Dennett, D.C.: 1987, *The Intentional Stance*, A Bradford Book, MIT Press, Cambridge, Mass./London, England.
13. Ellenberger, H.E.: 1970, *The Discovery of the Unconscious. The History and Evolution of Dynamic Psychiatry*, 2nd ed., Basic Books, Inc., Publishers, New York.
14. Engelhardt, D. von: 1988, 'Romanticism in Germany', in R. Porter and M. Teich (eds.), *Romanticism in National Context*, Cambridge University Press, Cambridge, pp. 109–133.
15. Fahrenberg, J.: 1979, 'Das Komplementaritätsprinzip in der psychophysiologischen Forschung und psychosomatischen Medizin', *Zeitschrift für Klinische Psychologie und Psychotherapie* **27**, 151–167.
16. Fechner, G.T.: 1851, *Zend-Avesta, oder über die Dinge des Himmels und des Jenseits*, 2 vols., Voss, Leipzig.
17. Fechner, G.T.: 1879, *Die Tagesansicht gegenüber der Nachtansicht*, Breitkopf und Hartel, Leipzig.
18. Freud, S. (and J. Breuer): 1972, 'Studien über Hysterie' (1895), in S. Freud, *Gesammelte Werke, Erster Band: Werke aus den Jahren 1892–1899*, 4th ed., S. Fischer Verlag, Frankfurt a.M., pp. 75–312.
19. Gruhle, H.W.: 1932, 'Geschichtliches', in [6], pp. 1–30.
20. Gruhle, H.W.: 1932, 'Die Psychopathologie' in [6], pp. 135–210.
21. Gusdorf, G.: 1984, *L'homme romantique (=Les sciences humaines et la pensée occidentale XI)*, Payot, Paris.
22. Hammer, F.: 1974, *Leib und Geschlecht. Philosophische Perspektive von Nietzsche bis Merleau-Ponty und phänomenologischer Aufriss*, Bouvier Verlag Herbert Grundmann, Bonn.
23. Hahn, P.: 1976, 'Die Entwicklung der psychosomatischen Medizin', in *Die Psychologie des 20. Jahrhunderts, Bd. 1: Die Europäische Tradition*, Kindler, Zürich, pp. 932–952.
24. Hahn, P. and Jacob, W. (eds.): 1987, *Viktor von Weizsäcker zum 100. Geburtstag*, Springer Verlag, Berlin/Heidelberg/New York.

25. Heidegger, M.: 1987, *Zollikoner Seminare. Protokolle-Gespräche-Briefe*, edited by Medard Boss, Vittorio Klostermann, Frankfurt a.M.
26. Heidelberger, M.: 1988, 'Fechners Leib-Seele-Theorie', in J. Brožek and H. Gundlach (eds.), *G.T. Fechner and Psychology. International Gustav Theodor Fechner Symposium, Passau, 12 to 14 June 1987* (=Passauer Schriften zur Psychologiegeschichte, Nr. 6), Passavia Universitätsverlag, Passau, pp. 61–77.
27. Henkelmann, T.: 1986, *Viktor von Weizsäcker (1886–1957). Materialien zu Leben und Werk*, Springer Verlag, Berlin/Heidelberg/New York.
28. Janzarik, W.: 1978, '100 Jahre Heidelberger Psychiatrie', *Heidelberger Jahrbücher XXII*, Universitätsgesellschaft Heidelberg, Heidelberg/Berlin/New York, pp. 93–113.
29. Jaspers, K.: 1959, *Allgemeine Psychopathologie*, 7th ed., Springer Verlag, Berlin/Göttingen/Heidelberg.
30. Jores, A.: 1950, *Vom Sinn der Krankheit*, Selbstverlag der Universität, Hamburg.
31. Jores, A.: 1970, *Der Mensch und seine Krankheit. Grundlagen einer anthropologischen Medizin*, 4th completely revised edition, Klett, Stuttgart.
32. Kasanmoentalib, S.: 1989, *De dans van dood en leven. De Gestaltkreis van Viktor von Weizsäcker in zijn wetenschapshistorische en filosofische context*, Kerckebosch, Zeist.
33. Kolle, K.: 1967, 'Zur Kritik der sogenannten Psychosomatik', in N. Petrilowitsch (ed.), *Zur Psychologie der Persönlichkeit*, Wissenschaftliche Buchgesellschaft, Darmstadt, pp. 168–194.
34. Kraepelin, E.: 1887, *Die Richtungen der psychiatrischen Forschung. Vortrag, gehalten bei Uebernahme des Lehramtes an der Kaiserlichen Universitat Dorpat*, Verlag von F.C.W. Vogel, Leipzig.
35. Krehl, L.: 1930, *Pathologische Physiologie*, 13th ed. (1st ed. 1893), F.C.W. Vogel, Leipzig.
36. Kütemeyer, M.: 1973, *Anthropologische Medizin oder die Entstehung einer neuen Wissenschaft. Zur Geschichte der Heidelberger Schule*, Diss. Heidelberg.
37. Kütemeyer, M.: 1981, 'Versuch der Integration psycho-somatischer Medizin in eine Neurologische Universitätsklinik', in Th. von Uexküll (ed.) *Integrierte Psychosomatische Medizin: Modelle in Praxis und Klinik*, F.K. Schattauer Verlag, Stuttgart/New York, pp. 187–226.
38. Laín Entralgo, P.: 1956, *Heilkunde in geschichtlicher Entscheidung. Einführung in die psychosomatische Pathologie*, O. Müller Verlag, Salzburg.
39. Lopez Ibor, J.J.: 1963, 'Psychosomatische Forschung', in H.W. Gruhle, R. Jung, W. Mayer-Gross and M. Müller (eds.), *Psychiatrie der Gegenwart, Bd. I/2: Grundlagen und Methoden der klinischen Psychiatrie*, Springer Verlag, Berlin/Göttingen/Heidelberg, pp. 77–133.
40. Luther, E.: 1967, *Historische und erkenntnistheoretische Wurzeln der medizinischen Anthropologie Viktor von Weizsäckers* (=Wissenschaftliche Beitrage der Martin-Luther-Universität, Halle/Wittenberg 1967/12 (R.5)), Halle (Saale).
41. Margetts, E.L.: 1954, 'Historical Notes on Psychosomatic Medicine', in [70], pp. 41–68.
42. Marquard, O.: 1987, *Transzendentaler Idealismus, Romantische Naturphilosophie, Psychoanalyse* (=Schriftenreihe zur philosophischen Praxis; Bd. 3),

Verlag für Philosophie Jürgen Dinter, Köln.
43. Mayr, C.F.: 1952 'Metaphysical Trends in Modern Pathology', *Bulletin of the History of Medicine* **26**, 71–81.
44. Nagel, T.: 1986, *The View from Nowhere*, Oxford University Press, New York/Oxford.
45. Reiner, H.D.: 1964, *Das Leib-Seele-Problem in der psychosomatischen Medizin bei Viktor von Weizsäcker*, Diss. München.
46. Rothschuh, K.E.: 1965, *Prinzipien der Medizin. Ein Wegweiser durch die Medizin*, Urban & Schwarzenberg, München/Berlin.
47. Rothschuh, K.E.: 1978, *Konzepte der Medizin in Vergangenheit und Gegenwart*, Hippokrates Verlag, Stuttgart.
48. Schipperges, H.: 1959, 'Leitlinien und Grenzen der Psychosomatik bei Friedrich Nasse', *Confinia Psychiatrica* **2**, 19–37.
49. Schipperges, H.: 1981, *Kosmos Anthropos: Entwürfe zu einer Philosophie des Leibes*, Klett-Cotta, Stuttgart.
50. Schmitz, H.: 1965, *System der Philosophie. Zweiter Band, Erster Teil: Der Leib*, H. Bouvier u.Co., Bonn.
51. Schonecke, O.W.: 1972, 'Wissenschaftstheoretische und methodologische Probleme der psychosomatischen Forschung und Theoriebildung', *Zeitschrift für Psychosomatischen Medizin und Psychoanalyse* **18**, 352–368.
52. Snell, B.: 1975, *Die Entdeckung des Geistes. Studien zur Entstehung des europäischen Denkens bei den Griechen*, 4th revised ed., Vandenhoeck & Ruprecht, Göttingen.
53. Strasser, S.: 1983, 'Das Problem der Leiblichkeit in der phänomenologischen Bewegung', in A.T. Tymienecka (ed.), *Analecta Husserliana*, vol. XVI, D. Reidel Publishing Co., Dordrecht, pp. 19–36.
54. Uexküll, Th. von: 1970, *Grundfragen der psychosomatischen Medizin*, 4th ed. (1st ed. 1963), Rohwolt Taschenbuch Verlag, Reinbek bei Hamburg.
55. Uexküll, Th. von (ed.): 1979, *Lehrbuch der Psychosomatischen Medizin*, Urban & Schwarzenberg, München/Wien/Baltimore.
56. Uexküll, Th. von (ed.): 1986, *Psychosomatische Medizin*, 3rd ed., Urban & Schwarzenberg, München/Wien/Baltimore.
57. Uexküll, Th. von and Wesiack, W.: 1979, 'Das Leib-Seele-Problem in psychosomatischer Sicht', in [53], pp. 56–71.
58. Verwey, G.: 1984, 'Antropologische geneeskunde in discussie', *Algemeen Nederlands Tijdschrift voor Wijsbegeerte* **76** (4), 207–227.
59. Verwey, G.: 1985, *Psychiatry in an Anthropological and Biomedical Context. Philosophical Presuppositions and Implications of German Psychiatry 1820–1870*, D. Reidel Publishing Co., Dordrecht/Boston/Lancaster.
60. Verwey, G.: 1988, 'Dualismen in de geschiedenis van de psychiatrie; een casus: de Heidelberger psychiatrie in het interbellum', in L. de Goei (ed.), *In de geest van het lichaam* (=NcGv-reeks 126), Nederlands centrum Geestelijke volksgezondheid, Utrecht, pp. 58–87.
61. Vesey, G.N.A.: 1968, 'Agent and Spectator. The Double-Aspect Theory', in G.N.A. Vesey (ed.), *The Human Agent*, Macmillan /St.Martin's Press, London/Melbourne/Toronto/New York, pp. 139–159.

62. Vesey, G.N.A. (ed.): 1964, *Body and Mind: Readings in Philosophy*, George Allen and Unwin Ltd., London.
63. Weitbrecht, H.J.: 1955, *Kritik der Psychosomatik*, Mit einem Geleitwort von Prof. Kurt Schneider, Georg Thieme Verlag, Stuttgart.
64. Weizsäcker, V. von: 1973, *Der Gestaltkreis* (1st ed. 1940), Suhrkamp Taschenbuch Verlag, Frankfurt a.M.
65. Weizsäcker, V. von: 1986ff, *Gesammelte Schriften*, 10 vols. (only vols. 1, 5–9 have appeared), Suhrkamp Verlag, Frankfurt a.M.
66. Weizsäcker, V. von: 1986, 'Einleitung zu G.Th. Fechner: Tages- und Nachtansicht' (1922), in *Gesammelte Schriften 1*, pp. 491–501.
67. Weizsäcker, V. von: 1987, 'Aertzliche Fragen. Vorlesungen über Allgemeine Therapie' (2nd revised ed., 1935), in *Gesammelte Schriften 5*, pp. 259–342.
68. Weizsäcker, V. von: 1986, 'Körpergeschehen und Neurose. Analytische Studie über somatische Symptombildungen' (1933), in *Gesammelte Schriften 6*, pp. 121–238.
69. Weizsäcker, V. von: 1987, 'Anonyma' (1946), in *Gesammelte Schriften 7*, pp. 43–89.
70. Weizsäcker, V. von: 1987, 'Meines Lebens hauptsächliches Bemühen' (1955), in *Gesammelte Schriften 7*, pp. 372–392.
71. Weizsäcker, V. von: 1988, 'Der kranke Mensch. Eine Einführung in die medizinische Anthropologie' (1951), in *Gesammelte Schriften 9*, pp. 325–641.
72. Weizsäcker, V. von and Wyss, D.: 1957, *Zwischen Medizin und Philosophie*, Vandenhoeck & Ruprecht, Göttingen.
73. Wenzel, A.: 1933, *Das Leib-Seele-Problem*, Felix Meiner Verlag, Leipzig.
74. Wittkower, E.D. and Cleghorn, R.A. (eds.): 1954, *Recent Developments in Psychosomatic Medicine*, Sir Isaac Pitman & Sons, London.
75. Wright, G.H. von: 1971, *Explanation and Understanding*, Routledge & Kegan Paul, London.
76. Wundt, W.: 1911, 'Ueber psychische Kausalität', in *Kleine Schriften, Zweiter Band*, Verlag von Wilhelm Engelmann, Leipzig, pp. 1–112.

STUART F. SPICKER

# INVULNERABILITY AND MEDICINE'S "PROMISE" OF IMMORTALITY: CHANGING IMAGES OF THE HUMAN BODY DURING THE GROWTH OF MEDICAL KNOWLEDGE

## CONTEMPORARY MEDICINE AND MEDICAL POWER

Historians in all likelihood would not easily reach a consensus if asked to locate temporally the transition to contemporary medicine, since various criteria could be selected in terms of which contemporary medicine (1900+) might be identified as having come into being (I am, of course, referring neither to "modern" nor to Oriental medicine). In order to avoid this debate, however interesting it may prove to be, and thereby to by-pass this certain challenge from medical historians, I simply confess my bias at the outset: the origin of contemporary medicine is typically and not inappropriately identified by the *general public* as having its origins in the discovery of new knowledge concerning the cellular and sub-cellular mechanisms that govern organic pathology, i.e., infection in the body, as well as even more recent knowledge of the appropriate use of particular instruments invented, developed, and refined for a wide variety of surgical and other technological interventions, which, it turns out, is virtually synonymous with the emergence of contemporary medical power – having passed through a very extended epoch of "medico-technical powerlessness" ["*medisch-technische onmacht*"] (Dutch [17], p.2). That is, Francis Bacon's maxim, "Knowledge is power" has long since 1620 been transformed from the pen of a British philosopher to contemporary science and technology, that includes virtually all medicinal and surgical interventions inconceivable by William Harvey in 1628, the year of publication of his monumental *Exercitatio anatomica de motu cordis et sanguinis in animalibus*.

In 1969, when the term 'bioethics' had not yet come into vogue, and the unfortunate title 'ethicist' had as yet not been coined (the OED simply defines it as "ethics" + "ist"), a renown Dutch psychiatrist and philosopher of medicine, Jan Hendrik van den Berg, published his *Medische macht en medische ethiek* [17]. In this extended essay van den Berg divided the history of medicine into three general epochs: (1) the long period of medical powerlessness – "from primeval times until 1870" (he apparently selected the 1870s because these were the years during which Koch and Pasteur proved that small living organisms cause infection); (2) the period of "transition from



*H.A.M.J. ten Have et al. (eds.), The Growth of Medical Knowledge, 163–175.*

technical powerlessness to technical power" – "from approximately 1870 until a few years ago" (that is, the late 60s); and (3) the era of "medico-technical power" – "the most recent years" since the late-60s ([17], p. 24).

I shall not dwell here, however, on the subtle aspects of the relation between medical *power* and medicine's centuries-old cryptic *knowledge*, since that could easily cause us to turn our attention to certain core concepts in the Foucauldean *corpus* (a complex subject, though one quite germane to this volume). Furthermore, I shall not pursue here any particular topic in the broad domain of bioethics, since this volume was designed to focus on the *epistemological* implications of the growth of *medical* knowledge, being the first publication to reflect the work of the European Society for Philosophy of Medicine and Health Care – a rather bold title, since a number of contemporary philosophers and physicians continue to claim that the use of the expression – "Philosophy *of* Medicine" – to flag a relatively recent addition to the domain of philosophy proper is merely putative, i.e., that the philosophy of medicine, unlike the philosophy of science or the philosophy of mind, for example, is not a true sub-discipline of philosophy [13]. Furthermore, perhaps one should remain troubled by a remark made some thirty-four years ago by Owsei Temkin, a historian of medicine, in his rarely cited 'On the Inter-relationship of the History and the Philosophy of Medicine'; there he made explicit reference to "the problematic nature of a philosophy of medicine" ([15], p. 245, note 10). One may notice too that historians of medicine like Temkin had been quite careful *not* to describe certain conceptual studies in medicine as studies in the philosophy *of* medicine. Although a glance at some earlier titles by twentieth-century historians of medicine reveals some awkwardness of expression, there was apparently no conceptual confusion or suggestion that philosophy *of* medicine was extant. For example, Temkin's *Galenism*, published in 1973, bears the subtitle *Rise and Decline of a Medical Philosophy* [16], and Walter Pagel's *Paracelsus* (1958) is subtitled *An Introduction to Philosophical Medicine in the Era of the Renaissance* [8]. The point I wish to make and to stress is that the preposition 'of' in the expression 'philosophy of medicine' signals not only the relatively late growth of a new limb on the *corpus philosophicum*, but is actually realized today through the formulation of and response to a host of new philosophical problems in medicine that reflect the appearance (perhaps rejuvenation) of an intellectual discipline that is, as perhaps more time will show, "the most important stimulus to philosophical reflection in the twentieth-century", a remark made seventeen years ago by Edmund D. Pellegrino ([9], p.20). It is important to recall that, in 1974, Dr.

Pellegrino was the first to offer the challenging suggestion that in the closing decades of the twentieth century "Medicine could provide the powerful stimulus to philosophy which Christian theology provided in the Middle Ages" ([9], p.20). With the establishment of this new sub-discipline of philosophy, given its own unique problems made further accessible to philosophical reflection and clarification, we would have at least partial affirmation that medicine is in fact *the* most important stimulus to philosophical reflection in our century, even if some are critical of that influence. But having asserted this, one must be careful to note that it was not the *possibility* of Christian theology, nor even the *practice* of Christianity (such as it was) that provided the impetus to philosophy in the Middle Ages, but Christian *theological speculation* among its more ardent and zealous practitioners, who were themselves exemplary thinkers. As with all speculative ideas, then, those that follow must move on their own if they are truly viable; I shall simply set them loose, and hope they will intrigue and not become sclerotic, atrophy, and die.

## THE GROWTH OF MEDICAL KNOWLEDGE AND THE ASSUMPTION OF INVULNERABILITY

Over a decade ago, I published an essay in *The Journal of Medicine and Philosophy* on the theme "Idea and Image of Man" in contemporary medicine [12]. In exploring this theme I noted that it was necessary to "attend to various images of man especially those images which have served to provide key insights into man's nature or 'condition,' the latter term preferred by Continental philosophical anthropologists" ([12], p. 105). Then I asked: "Is another image of man waiting in the wings? Indeed, is it already with us in the effigy of 'total man,' 'future man,' 'healthy man,' or (since man as microcosm is merely a relic of Renaissance reverie) in the form of 'microscopic' or 'molecular man,' the other end of the macrocosm?" ([12], p.105). Had I been more thorough then, I would have added 'genetic man' – humanity conceived as essentially composed of nucleic acids, i.e., DNA. Even nonbiologists are familiar with this effigy and refer to it obliquely in various quips, like Samuel Butler's famous aphorism of 1877 – "A hen is only an egg's way of making another egg" – or, in more modern parlance – "A human being is only DNA's way of making more DNA." In that essay of 1976 I did my best to avoid the trap that yawns for those who purport to have discovered *the* image or essence of man, and thus I made reference to Max Scheler's philosophical anthropology in which he differentiated between (1) *homo religiosus*, man of faith in divinity, (2) *homo sapiens*, "rational man," which later became bifurcated into "rational man" and "evolutionary man," (3) *homo faber*, man as tool user, not qualitatively different from other animals, (4) *homo Dionysiacus*, where reason is the infirmity of life and decadence is celebrated, and finally (5) *homo creator*, man as culture-producing actor ([10], ch.IV; [11]). I concluded with the modest thesis that human *infirmity*, and especially modern medicine's ability to cure, generated yet another image of man appropriate to our time, but one which should not be adopted as *the* image of man for all time, but rather one which merely complements other conceptions by placing special emphasis on *homo sapiens* as *infirma species*, as 'fallen man' – infirmity, instability, and *vulnerability* being essential to any comprehensive understanding of the human condition. Yet as we write this image is already undergoing transformation. Let me explain.

Dialectical, if not diabolical, tacit tendencies serve to foreshadow a novel and yet prevailing image of humanity in the context of modern medicine and medical power: On the one extreme North Americans and Europeans, given their overconfidence in the "miracles" of modern medicine, tend to view themselves as invulnerable, virile, and, even when seriously ill, retrievable from death's grasp (surely a rather special mode of "immortality"). This is so in part because we have generally come to enjoy a heretofore unanticipated yet significant increase in average life expectancy. Moreover, we are psychically unprepared for immanent decline, disability, disease, and even death; that is, there prevails the Janus face of being saved from death *now*, though we may (as a merely theoretical possibility) die of something *tomorrow* – but we shall surely not die now!

On the other hand, there had been an earlier image of humanity as vulnerable, mortal, finite, and even fallible. But this no longer prevails. Today's tacit image of invulnerability, if not an image of full-fledged immortality, now receives its formal expression in various cultural phenomena like (1) 'cryonics' (coined in 1965) – the process of using ethylene-glycol to freeze a "dead" diseased human in the hope that in the interim biomedical research will develop a cure for the disease so in time the persons can be brought back to life which was simply interrupted. Once a cure is discovered, the cryonic contract requires that the body be thawed and the past person restored to life. [Could this only happen in California, where bizarre social events, like the recent celebration of the 20th anniversary of the American Cryonics Society (*The Times–New York*, Jan. 19, 1989), convene and cryptic captions serve to entice?: "Many are cold but few are frozen" (a quip I owe to Stanley Reiser]; (2) In addition, physicians themselves are encouraging the public to think in terms of living a life approaching 200 years – this is no mere extension of average life expectancy, but rather a radical increase in longevity. How different is this image from religion's promise of immortal life? If one can not quite imagine 200 years of life, one can at least understand WHO's recent battlecry: "Health for all by the year 2000!" How different is this image from religion's promise of immortal life? This should give us pause....

The promise of an immortal life, made in earnest and grounded in religious authority, is of course the province of established religion. Although the secretiveness of the medical profession in antiquity derived from religious motives, traceable perhaps to the fourth and third centuries B.C. (and irrespective of the fact that whatever secretiveness remains to-day is more likely due to the desire of physicians to reduce competition of a saleable good than to honor the holy purity of a religious or even cult community), to-day's medical "silence" is intrinsically connected to the promise of a dramatically extended longevity. Immortal life was religion's mystery never Western medicine's promise. With the transition to the epoch of medical power in the twentieth century, and the partial if not complete "eclipse of revelation" in our time (as Hans Jonas has observed), ([6], p. xvi), H.T. Engelhardt can, I believe, quite convincingly claim, that whereas "The Gods once brought rain and cured diseases, [N]ow there is meteorology and pharmacology. The presence of the divine has been exorcised.... We live on the other side of a religious age" ([3], pp. 79–80). The promise of religion should not of course have become the hope of medicine – medicine might at times make an outrageous claim, but certainly remained prudent by avoiding the arrogance associated with the hope and promise of eternal life. But medicine's prudence is clearly not the public's measure: Witness the assumed invulnerability of our earthly embodiment, the turn to cryonics and the secular faith in biomedical knowledge eventually capable of decoding the mystery of and intervening to reset the "genetic clock" that governs at the sub-cellular level all life processes, and the unwarranted optimism of a number of medical professionals that we can extend the longevity of our species to almost 200 years. Indeed, these elements have all converged to alter our view of physicians, and to encourage us unjustifiably to expect that they reward us with quasi-immortal life. Paradoxically, then, while we discover that the collapse of belief in religion's ancient promise of immortality has enhanced the image of the physician and his medical power, we find it has resulted in much greater difficulty in the clinic where the attending physician must make a myriad of medical and ethical decisions. For the physician at the bedside of the dying can no longer simply elect to ease his patient's passing (surely he can postpone his patient's death *now*, which is momentarily as good as immortality), and yet he can not simply prolong his patient's life indefinitely, though he often has the technical means to do so (which for many is as good as immortality now, the step just prior to cryonic "preservation"). After all, as physicians well appreciate, the public continues to perceive the neo-cortically brain dead being as a potentially viable person, still present, still living, notwithstanding the medical consensus that the personhood of the patient is no longer available and will not return; for in these cases only the "neomort" is truly present – the respiring corpse, in limbo, which merely offers the appearance of viable personal life, when in fact the respirator is the truly efficient cause,"breathing" on behalf of the neomort. Such phenomena signal that notwithstanding this contemporary era of medical power, the traditional goals of medicine are still with us. As John Gregory observed in 1770, the trifold purpose of medicine is to (1) preserve health, (2) cure disease, and (3) prolong life ([5], Preface). It is the last of these that has recently undergone the most radical transformation. But the truth is that we have now inappropriately (foolishly?) come to expect the impossible from medicine. The *idée fixe* prevails: that we shall always be cured or at least have our condition arrested and that the next miracle cure or drug for the myriad of our ills is just around the bend. But that dream image is presently being shattered, though reconstituted and made corrigible momentarily by cryptic remarks to the media. [The former U.S. Surgeon General, Dr. C. Everett Koop, had publicly declared that a conservative date for the discovery of a vaccine for the cure of AIDS and the prevention of HIV sero-conversion will be the late 1990s.] Exaggerating the spirit of such optimism we find that some women no longer expect to experience pain – not even in childbirth. We demand to die peacefully, painlessly, "naturally," nay, that we not die at all! In sum, medicine is to fulfil what some claim it to possess as its natural or tacit promise: the promise of freedom from death itself. We are never to leave the circle of life which began in health, passed through periods of infirmity which (especially if we are fortunate in the "natural" and "social lotteries") elicited the care of professionals, and led to rehabilitation, recuperation, health, and wholeness – but no death. Ironically, this image is also shared by persons within medicine, and forces us to ask: Do we to-day unconsciously share this dream image? If we do, is it because we have failed to acknowledge the intrinsic fallibility of empirical knowledge, especially medical knowledge, whose phenomenal growth since 1870 is unquestionable? (I leave aside here the claim that we have a superb understanding, e.g., of human anatomy, and that nothing more can be learned in this domain.) Have we failed to appreciate the indisputable fact that all empirical, biomedical, and clinical knowledge (though the outcome of humanity's continuous quest for certainty) is intrinsically provisional, probabilistic, incomplete, and subject to reinterpretation if not outright refutation by means of experimental methods of verification, where unfortunately the prevailing "wisdom" rarely if ever reflects a challenge regarding the foundations of public confidence?

Perhaps phenomena like the current AIDS epidemic – though initially projected as a world-wide catastrophe around 1981, and presently challenging the best minds in biomedicine who seek a cure or at least dramatic and foolproof methods of prevention – will at last challenge contemporary humanity's assumed invulnerability. This is now being confronted by the insidious lethality of AIDS, the uncertainties about its spread, and the total absence of any effective treatment – all of which threaten our own personal safety, and especially the safety of health professionals. Are we in existential conflict with our previous presumption of our bodily invulnerability upon which we have unjustifiably come to rely? Now that ancient Hippocratic medicine's secretiveness is being replaced by openness to rational analysis and criticism, as well as open debate, where most health professionals agree that the competent patient has a right to access and to share in the medical knowledge and decisions germane to his care (however reexpressed for the lay persons' understanding, and especially since the acquisition of new knowledge is supported by public funds, not to mention the public's desire to eliminate what it often misconstrues as medical paternalism), are we now prepared to acknowledge our intrinsic vulnerability, fallibility, openness to infirmity, and essential mortality? The answer rests in great measure with our self-understanding and the image of man Scheler called '*homo religiosus*', as well as the more recent transformation of contemporary humanity's conception of its embodiment.

## *HOMO RELIGIOSUS* AND THE IMAGE OF THE HUMAN BODY

A very erudite scholar opened his lecture on the Middle Ages with the quip that "The Middle Ages are not what they used to be"; after all, reconstructing human history is always a risky enterprise. Be that as it may, it seems safe to say that during that long but not so dark period of Occidental history, the image of *homo religiosus* dominated the belief systems, thoughts, and activities of mankind. During those epochs, when the image of *homo religiosus* prevailed – and the medieval image (or better, "vision") of mankind always included the notions of birth, life, aging, and death – humanity's infirmities, i.e., illness, disease, pain, suffering, and permanent disability were given their lowly "place" in the metaphysical hierarchy or Great Chain of Being, though still attended to in the numerous medical writings of physicians.

In his defence of theory in medicine, a philosopher, Scott Buchanan, expressed the myth that goes by the name "the doctrine of signatures," that has served in a very general way to provide a metaphorical account of the ability to discover remedies for all forms of animal and human sickness. "God created man and made him subject to ills and misfortunes. He also created stones, plants, animals and other human beings and marked them with characteristic marks to be recognized and used as remedies. He gave man the ability to read these signatures both in himself and in natural objects and to interpret them for the good of his soul and body" ([2], p. 33).

This doctrine is deeply connected to the vision of man as *homo religiosus*, and serves in part to account for the connection between mankind's Fall from Divine grace and his mortality in the absence of that grace. For after the Fall mankind was also condemned to a "medical" Garden of Eden, a garden no longer free of infirmity, pain, suffering, and death. Not wishing to push this Biblical image too far, still the general implication for us is clear: infirmity as an essential condition of our human embodiment was not until quite recently considered too seriously, for it did not, as a condition of being human, rank among the most focal aspects of our finitude. From our twentieth-century perspective, however, *homo religiosus* no longer satisfies the human spirit, and this conception or image clearly no longer functions as the cryptic image for contemporary humanity. But having said this we should not yet retreat from biblical myth: Perhaps we can penetrate to the *Ursprung* of our current tendency to neglect our essential vulnerability, infirmity, our openness to disease, illness, pain, suffering, and death. A clue can be found in the reflections on the human body found in the writings of a physician who lived in the early sixteenth-century – Theophrastus Bombastus von Hohenheim, or Philippus von Hohenheim, nicknamed Aureolus, namely, Paracelsus (1493/4–1541).

## THE SIXTEENTH-CENTURY MEDICAL IMAGE OF THE HUMAN BODY

In an interesting passage in "Paracelse," Alexandre Koyré observes that before the Fall the body of Adam was a dynamic body; it did not nourish itself as our's does; it did not eat [7]. For though all living beings require nourishment, Paracelsus held, superior beings (angels, for example), did not need to feed themselves by mouth as animals do and as we have done since. These beings did not live in bodies like ours.

All created beings were not in themselves in need of a rapid force to replace what they consumed and metabolized; they were not in need of any restorative influence. In short, before the Fall, the body of Adam was *not* an organic body; it did not metabolize. But, we might ask, what kind of living body is it that does not eat, digest, or die? It certainly is not a body open to infirmity which thereby would provide the *raison d'être* of medicine. Before the Fall, Adam's body was (to use one of Paracelsus's "monsters of terminology" [14]), an "iliastric" body, the body that contemplates eternity and the higher sphere of the Creator. After the Fall, the human body is a "cagastric" body, the body that issued from the person of Eve, like the earth, the material universe – temporal, mutable, and corruptible ([7], pp. 106–107). [Koyré observes that in 16th-century representations of the human face, the left eye was frequently omitted, since it symbolized the earthly, temporal, "cagastric" eye that was to be kept closed; the right eye, the "iliastric," contemplates eternity and is therefore represented ([7], p. 106; [8], p. 235, fig. 23). Thus we note here the image of man as microcosmos: man reflecting the image of the world, the world the image of God, and man the image of God, as if by force of a transitive cosmic relation.]

Whereas Galileo Galilei understood himself to be reading the book of Nature, whose language was that of mathematical symbols and notation, Paracelsus was "reading the signs" of the cagastric body, his task being to decipher the secrets of the human body by seeking to understanding the language in which its signatures were written. Whereas Galileo understood the language of Nature (natural philosophy) to be mathematical symbolization, Paracelsus in truth had no language for comprehending the cagastric, infirm, or diseased body. The closest language was to be derived from the patient's urine, or other similar signatures "enscribed" in other animals and plants. In fact, it was precisely the lack of preexisting terms and a proper medical language or nosology of disease that motivated the creation of such terms as "cagastric" and "iliastric" body ([14], p. 216).

In Paracelsus' writings, three levels of material organization can be distinguished: (1) the matter of *terra firma*, the firmament, palpable or gross matter; (2) the matter of the living, human body; and, (3) astral matter, the matter of the stars and divine bodies. The first two generate the category of the cagastric body; the latter the iliastric. The Renaissance image of the human body, the one to which in its concrete form medicine attends is, in the language of Paracelsus, the cagastric body, the product of a Fall – perverse, corrupt, and base. Perfect or divine bodies are not exemplified by human bodies, but by metals, especially gold. [Once one appreciates the function of the tincture, the analogue for Christ's astral or iliastric body, one has an important key to Paracelsian medicine.] The cryptic image of man as *homo religiosus* is thus connected conceptually with the iliastric not the cagastric body, the latter being found in the alchemists' crucible, and the physician's urine flask. In the writings of Paracelsus, then, we find a double perspective – at least two categories of body – which for us to-day is reduced to a myopic or Cartesian one, the "cagastric", the physiological organism, the anatomical, corruptible body, devoid of magic and possessing no vital force; there is, it appears, only the organic body and its eventual inorganic status, the corpse (or perhaps W. Gaylin's "neomort" prior to the corpse), which decomposes in the earth or tomb. The disappearance of the iliastric, astral body is at one with the eclipse of revelation, and has accompanied the gradual vitiation and enervation of religious authority. For cagastric bodies cannot be "saved," and if they are not saved, we are not saved. Recall that in his commentary on I Corinthians 15, Aquinas says that "my soul is not I; and if only souls are saved, *I* am not saved, nor is any man."

Reduced to a cagastric body, the living, metabolizing human body is the most subtle of natural alchemists, which even evacuates its own wastes. In nourishing ourselves *we* are the natural alchemists. Nutriment is reconstructed and we extract what is required for our sustenance. From the standpoint of the cagastric body illness is nothing in itself; it can only be understood within the full context of the life of each living being in its complex relations with the organic world; illness emerges according to its own nature, and thus for Paracelsus it does no good merely to treat symptom-signatures. The medicine of Paracelsus is not, therefore, a medicine of symptoms, but a medicine of *causes*; for example, illness may be caused by the corruption of gross matter, hostile astral forces, sources within the organism, or by the corruption of the personal self. Any and all of these causes are contingent. Under Paracelsus' vision, infirmity is fundamentally accidental, due to forces over which we generally have no control or suasion. But perhaps infirmity itself is of the very morphology of man's essential individuality and marks the essential difference between *homo sapiens* and all other animal species. J.C.F. Hölderlin muses: "...ihn scheun die Tiere, denn ein anderer ist, wie sie Der Mensch." ["...the animals shun him, for different from them is man."] (1798).

## CONTEMPORARY MEDICINE AND QUASI-IMMORTALITY

With the emergence of secular pluralist societies, *homo religiosus* began to vanish, the notion of "eclipse" being too generous, because the darkness of an eclipse is only *temporary*. Contemporary medicine, founded on scientific principles and the canons of inductive reasoning, has brought about a transition from the image of the human body as base, perverse, and corrupt, to one already found in Aristotle's writings – in his notions of "*dynamis*" and "*energeia*," which we have partially retained in such concepts as the body's "metabolic processes," its "homeostatic," "immunologic" and "self-reparation" mechanisms, through which the organism is magnificently "designed" to sustain itself – notwithstanding the anathema of the recent emergence of HIV sero-positivity and the acquired immuno-deficiency syndrome.

In sum, powerful religious lenses that distorted the human body, denigrating it as corrupt not only in sickness but even in health, were eventually replaced by our century – we now celebrate *homo Dionysiacus* and the lived body. This is analogous to the way religious authority in moral matters has given way to secular pluralistic values that thrive in peaceable societies where the joy of reasonable conversation and negotiation prevail among differing and radically divergent moral points of view [3].

## ADMONITION: MEDICAL ARROGANCE AND THE GROWTH OF MEDICAL KNOWLEDGE

In my essay of 1976 I left Dr. Pellegrino's claim unaddressed – that medicine will provide the most important stimulus to contemporary philosophy that Christian theology provided in the Middle Ages. At that time I concluded that his thesis remained to be proven. In retrospect, some fourteen years later, I believe his insight was quite prescient. Perhaps not too many decades from now historians of philosophy and medicine will attest to the truth of his judgment; we can only wait patiently. Given the extant philosophical literature in North America and Europe, I am inclined to wager that no human enterprise – art, natural or social science, technology in all of its manifestations, including cryonics – will have challenged the human understanding and philosophy itself in the twentieth century more powerfully than the ever growing body of knowledge derived from anatomy, physiology, pathology, immunology, clinical medicine, the various biomedical sciences, and the careful reflection on this knowledge by contemporary philosophers of medicine [1].

Even more importantly, we must continuously remain alert to a new and even more worrisome sense of *medische macht*, not simply the authority and power of physicians to impose their will in order to cure their patients or to forestall death by prolonging their lives (even at times doing moral injury by improperly prolonging dying) – but the misguided view that physicians and medicine broadly construed should seek to achieve sufficient knowledge, power, and prestige in order eventually to justify filling the lacuna created by the on-going collapse of traditional religious authority. We must, in short, remain vigilant and guard against an imperious arrogance: that the profession of medicine may be tempted to substitute itself for religious authority and once again proffer the false promise of quasi-immortal life – surely among the most powerful and destructive ideas capable of claiming the allegiance of those who already seek to undermine a society that desires to survive in peace.

## AFTERWORD

In 1976, I opened my essay citing the hopeful words of Goethe, penned in 1814. As an apology to those whom I may have offended unintentionally, permit me to close with Goethe's words, leaving in peace all those who hope for the evermore:

> Und so lang du das nicht hast,
> Dieses: Stirb und werde!
> Bist du nur ein trüber Gast
> Auf der dunklen Erde [4].
> [And as long as you don't possess it,
> This: Die and become!
> You are only a sad guest
> On this darkening earth.]

*School of Medicine*
*University of Connecticut Health Center*
*Farmington, Connecticut, U.S.A.*

## ACKNOWLEDGEMENT

I am grateful to my colleagues, Doctors Robert U. Massey and Henk A.M.J. ten Have, for taking the time to provide a careful critique of and to advance important suggestions for improving the penultimate manuscript.

## BIBLIOGRAPHY


1. Bondeson, W. B., Engelhardt, H. T., Spicker S. F., White J. M. (eds.): 1982, *New Knowledge in the Biomedical Sciences.* (*Philosophy and Medicine,* Vol. 10), D. Reidel Publishing Co., Dordrecht, Holland/Boston, U.S.A.
2. Buchanan, S.: 1938, *The Doctrine of Signatures: A Defense of Theory in Medicine.* Kegan Paul, Trench & Trubner, London.
3. Engelhardt, H. T., Jr.: 1985, 'Looking for God and Finding the Abyss: Bioethics and Natural Theology', in E. E. Shelp (ed.), *Theology and Bioethics: Exploring the Foundations and Frontiers.* D. Reidel Publishing Co., Dordrecht, Holland/Boston, U.S.A., pp. 79–91.
4. Goethe, W.: 1814 (July 31): *Der West-oestliche Divan*, Buch des Sängers. Selige, Sehnsucht.
5. Gregory, J. L.: 1770, *Observations on the Duties and Offices of a Physician, and on the Method of Prosecuting Enquiries in Philosophy.* W. Strahan & T. Cadell, London.
6. Jonas, H.: 1974, *Philosophical Essays: From Ancient Creed to Technological Man.* Prentice Hall, Inc., Englewood Cliffs, New Jersey.
7. Koyré, A.: 1971, *Mystiques, spirituels, alchimistes du xvi* e siècle allemand. Editions Gallimard, Paris, especially 'Paracelse', Ch. 3, pp. 75–129.
8. Pagel, W.: 1958, *Paracelsus: An Introduction to Philosophical Medicine in the era of the Renaissance.* S. Karger, Basel.
9. Pellegrino, E. D.: 1974, 'Medicine and Philosophy – Some Notes on the Flirtation of Minerva and Aesculapius', Annual Oration of the Society for Health and Human Values, Nov. 8, 1973, Washington, D.C. (Published by the Society in Philadelphia).
10. Scheler, M.: 1958, *Philosophical Perspectives.* trans. O. A. Haac (original text, 1954) Beacon Press; Boston, U.S.A., especially chapter IV.
11. Scheler, M.: 1961, *Man's Place in Nature.* trans. H. Meyerhoff (original text, 1928), Noonday Press, Division of Farrar, Strauss and Cudahy, New York.
12. Spicker, S. F.: 1976, '*Terra Firma* and Infirma Species: From Medical Philosophical Anthropology to Philosophy of Medicine', *Journal of Medicine Philosophy* **1**(2), 104–135. [In "Idea and Image of Man"]
13. Spicker, S.F.: 1986, 'Cognitive and Conative Issues in Contemporary Philosophy of Medicine', *The Journal of Medicine and Philosophy* **11**(1), 107–117.
14. Temkin, O.: 1952, 'The Elusiveness of Paracelsus', *Bulletin of the History of Medicine* **26**(3), 215–216.
15. Temkin, O.: 1956, 'On the Interrelationship of the History and the Philosophy of Medicine', *Bulletin of the History of Medicine* **30**(3), 241–251.
16. Temkin, O.: 1973, *Galenism: Rise and Decline of a Medical Philosophy*, Cornell University Press, Ithaca, New York.
17. van den Berg, J.H.: 1969, *Medische macht en medische ethiek.* Uitgeverij G. F. Callenbach N.V. Nijkerk, Netherlands; 1978, trans. and revised, *Medical Power and Medical Ethics*, W. W. Norton & Co, New York.

UFFE JUUL JENSEN

# VALUES AND THE GROWTH OF MEDICAL KNOWLEDGE

Does the growth of medical knowledge lead to a gradual overcoming of our vulnerability and infirmity, or has 'scientific' medicine obtained resources for its own development and secured its own power and authority in modern society by spreading the myth that it could make us transcend human weakness and pave the way to human immortality? Advocates of modern medicine take the first approach. Critics, however, consider medicine as one of the most ideology-producing enterprises of modern society.

It is a great merit of Stuart Spicker's essay that it shows that things are a bit more complicated than they appear in public debates.

In a very illuminating way Spicker traces the origin of our present tendency to neglect our own vulnerability, our openness to disease and death. He points out that this is not a myth produced by modern biomedicine, but an attitude also widespread in pre-modern medicine, e.g., in the thinking of Paracelsus. In Paracelsian medicine all causes of diseases (corruption of gross-matter, hostile astral forces, or sources within the organism or the self) are, as stressed by Spicker, contingent; so, infirmity does not *essentially* belong to our nature.

What Spicker points out is certainly not of historical interest alone. The Paracelsian ideas are, on the contrary, widely echoed by many contemporary spokesmen of holistic medicine who try to provide alternatives to the dominating biomedical approach of modern health-care systems. Holistic medicine does not embody one particular theory or specific therapeutic procedure – quite to the contrary. But at least a great number of holistic approaches seem to share some core assumptions concerning the nature of health and disease, disease being conceived as an inbalance, lack of harmony, or deviation from some original or natural state of health. Therefore biomedicine is criticized for not taking the natural resources of the body (or the environment) into account in therapy.

With due respect to the important differences between modern biomedical thinking and its holistic rivals they seem to share the basic idea of diseases having *specific* causes; indeed they share a therapeutic optimism related to that idea – that when the causes are found and eliminated the patient will regain his or her natural, normal, or healthy state. The two positions "simply"



*H.A.M.J. ten Have et al. (eds.), The Growth of Medical Knowledge, 177–185.*

disagree about how to characterize or define the state of health or balance in terms of specific biological theories (e.g., genetics) or metaphysical ones.

These surprising similarities between holistic medicine (and its pre-modern ancestors) and trends in modern biomedical thinking are, however, not traced further by Spicker. Though this would be a fruitful startingpoint for analyzing the ongoing influence of the image of invulnerable man, Spicker primarily focuses on differences between modern medicine and its pre-modern predecessors.

According to Spicker, the emergence of secular pluralist societies and scientific medicine "has brought about a transition from the image of the human body as base, perverse, and corrupt, to one already found in Aristotle's writings – in his notions of '*dynamis*' and '*energeia*', which we have partially retained in such concepts as the body's "metabolic processes" ... through which the organism is magnificently 'designed' to sustain itself ...." ([5], p. 173).

But stressing these aspects of the transition from medical thinking that embodies an ideal of *homo religiosus* to contemporary medicine, Spicker seems to ignore the important similarities pointed out above.

In the passage just quoted Spicker stresses certain aspects of the pre-modern view of man (or rather man's body). But this image is more complex than indicated in the characterization above. In the Paracelsian perspective, our body is not just corrupt; it is at least not *essentially* corrupt. Balance and harmony can be regained, if we only knew the secret of the vital force (Achaeus) [1].

Behind Spicker's account of the transition from Paracelsian to modern medicine one feels a certain ambivalence towards modern medicine (an ambivalence which many of us may share). On the one hand, Spicker addresses himself to a "promise of immortality" and an image of invulnerable man tacit in modern medicine; on the other hand, he honors modern medicine and its critical methods as a basic intellectual resource for settling with old myths and images.

Spicker apparently "represses" these mixed or contradictory feelings toward modern medicine, and transforms them to a theory about contradictionary tendencies in contemporary *society*: on the one hand we find, according to Spicker, a widespread ideology expressing the tacit promise of freedom from death, a kind of foundationalism pointing to salvation through scientific and technological progress; on the other hand there is a critical medical science (and practice) aware of its own limitations and of the uncertainty of medical decisions.

Where do we find the roots of the ideology of medico-technological optimism? This is only vaguely answered by Spicker. In one context he speaks of "North Americans and Europeans" who, "given their overconfidence in the 'miracles' of modern medicine, tend to view themselves as invulnerable, virile, and, even when seriously ill, retrievable from death's grasp ...." ([5], p. 166). It is not quite clear who Spicker, more exactly, is referring to, but it is probably an ideal type of 'the modern man', "unprepared for immanent decline, disease and even death".

However, though Spicker dissociates this ideology from modern medical thinking, individual physicians or groups of physicians are held responsible for taking part in creating this vain optimism concerning human omnipotence – "physicians themselves are encouraging the public to think in terms of living a life approaching 200 years ...." ([5], p. 167). Several other hints indicate that modern man's unwillingness to face his vulnerability and mortality cannot be interpreted as just an impersonal modern *Zeitgeist* nourished by the anxiety and Godlessness (forsakenness) of modern man. No, modern medicine (and philosophies associated with it) play a major role in the construction and spreading of the image of invulnerable man.

This implies that modern medicine itself should be scrutinized carefully to locate the elements or aspects of medicine which are to blame – and if possible to be changed. It is as if Spicker refrains from taking such a critical stance, except in very general terms toward the end of his essay, warning us that "the profession of medicine may be tempted to substitute itself for religious authority and once again proffer the false promise of quasi-immortal life ..." ([5], p. 174).

But who is to be blamed within this immense structure called modern medicine which embodies technical skills, scientific expertise, economic interest, personal and collective power? Perhaps these values governing modern medicine where it dominates specialized clinical treatment: prolongation of human life (a practice which I elsewhere [1] have labelled "the disease-oriented practice").

## ETHICAL ASPECTS OF 'PROLONGATION OF LIFE'

To-day we may perhaps wonder how the ideal of 'life prolongation' became dominant in medicine and how medical progress was (and is) measured by reference to the yardstick of survival (as embodied in the 'five-year survival test' being used to assess the quality of competing therapeutic strategies). If we wonder why things are so, it shows how short our historical memory is. It is not more than 60 years ago that death was still an ever-present threat even at moments where life was most strongly experienced: Mothers and their newborn babies were dying in great number, strong and vigorous men and women were struck by diseases and died without doctors being able to do anything. Even around the thirties, Lewis Thomas reminds us, medical students "during the third and fourth years of school began to learn something that worried us all, although it was not much talked about ... it gradually dawned on us that we didn't know much that was really useful, that we could do nothing to change the course of the great majority of the diseases we were so busy analyzing, that medicine, for all its facade as a learned profession, was in real life a profoundly ignorant profession" ([7], p. 29); "ignorant" here means lacking the power to realize its own end, to fight the ever-present death.

It would be a great mistake to think that it was the medical profession which imposed its professional ideals on laymen. The ideal to fight death and give mothers, children, poor people – actually all of us – the opportunity to live just a bit longer than granted by destiny seems to be an ideal of modernity and now shared by professionals and laymen alike.

And it is not difficult to understand the euphoria of doctors and patients when sulfanilamide heralded a revolution in medicine around 1937. Thomas recalls the experience: "I remember the astonishment when the first cases of pneumococcal and streptococcal septicemia were treated in Boston in 1937. The phenomenon was almost beyond belief. Here were moribund patients, who would surely have died without treatment, improving in their appearance within a matter of hours of being given the medicine and feeling entirely well within the next day or so" ([7], p. 35).

Prolongation of life is the basic value or ideal of the disease-oriented practice, but this internal-practice value only came to play a dominant role in modern society because it was in harmony with or could be justified by reference to a widely accepted ethical value not bound, I think, to any particular practice but rooted in everyday social practice.

It is part of our ordinary experience that we are dependent upon each other (our parents, relatives, colleagues, professionals). We are not free-floating atoms but subject to constraints of nature and of society. This fact (that we are in the power of circumstances and other humans) signals that moral considerations are a part of ordinary thinking. We appeal to moral responsibility when individuals or groups are subordinated to conditions which make them weak and restrain their possibilities for self-maintenance. Our model for morally responsible use of power can be summed up as follows:

when A causes B to do something which he would not have done otherwise, then A is acting morally responsibly if his actions towards A aim at or contribute to making B independent of the power of A; or others having similar power.

Throughout the history of modern society, science and technology have developed as part of communal efforts to control natural and social conditions, restraining our possibilities for action. Under conditions where death was an ever-present threat, fighting death became just one among many attempts to control natural and social conditions. Medical professionals were acting responsibly – in our quite ordinary understanding of that term – in developing the procedures of disease-oriented practice, but *moral* only under one specific condition: that the actions of medical professionals aim at or contribute to making patients *independent* of medical or other professional power.

And isn't this exactly the reason why so many react against modern biomedicine: that its procedures and strategies (though contributing to a dramatic increase in average life-expectancy which is in itself a debatable claim) demand a high price for its apparent success; that we become unceasingly dependent on medicine and its power, losing our ability for self-maintenance?

What does all this imply concerning our idea of the growth of medical knowledge? That we in the end become victims of growing medical knowledge? That knowledge should be held in check by emotions *and* ethics? No, this would be to accept a dichotomy between science and values deeply rooted in our culture and theoretically sanctioned by logical positivism.

In opposition to that it should be argued that no science or practice can determine its own standard of growth and progress independent of our social world and life world.

Any practice has its internal standards of quality. But aiming at growth and progress the practice has to be assessed from outside. As there is no absolute foundation, no God's-eye perspective from which such assessment could be carried out, we have to seek such external standpoints in our culture and history and in ordinary social life (I have argued for this elsewhere [3]).

I have already suggested that a model of responsibility rooted in everyday life has been the external, ethical standard for assessing medicine – and so also an implicit standard of growth of medical knowledge throughout the history of medicine. And I advocate a continuous use of that standard or value, presupposing that it is still the prototype of responsibility of ordinary life.

It is highly probable that physicians fighting human vulnerability with the heavy weapons of biomedicine conceive of their own therapeutic practice as realizing human self-maintenance as implied by the principle of responsibility I sketched above. If in spite of that it appears that modern biomedical strategies – at least to some degree – lead to human dependency, then we face a very serious "illness": *professional self-deception* ("self-deception" defined as a process of individuals or collectives undermining ends and values they themselves espouse [2]).

If the basic values of modern medicine are sound, i.e., if they are in agreement with our ordinary moral intuitions (e.g., that we are always dependent upon each other, and that the weak should be supported to secure his or her self-maintenance as far as possible); if the probabilistic (and anti-foundationalist) methods of modern disease-oriented practice contain a vaccine against omnipotent fantasies, then we have to look elsewhere in the practice and institutions of modern medicine to explain how it continues to spread the ideology of invulnerable man.

One must call for concrete analyses of the practice of medicine, of economic interests influencing the development of biomedicine and medical technology, as well as of the way medical students and young physicians are socialized and adapt to a hierarchical and competitive system worshiping the powerful, self-sacrificing, invulnerable physician as their idol.

If a democratized and decentralized health care system (e.g., as suggested by WHO in *Health for all by the year 2000*) could help to destroy that idol and confront the fear of weakness and repression of vulnerability so typical of modern society, perhaps that too should be advocated.

## PSYCHO-SOMATICS AND GROWTH OF MEDICAL KNOWLEDGE

The very idea of the 'growth of medical knowledge' is usually associated with the rise and development of biomedicine. So, Verwey's claim, that psychosomatics, by insisting upon "the unsolvable riddle of the mind-body relationship, which is the sphinx of our own human psychosomatic existence" has contributed to the growth of knowledge, is most provocative [8].

Verwey's position raises various questions: How can perennial philosophical riddles or problems contribute to the growth of scientific knowledge? Should we not rather find new ways to overcome these philosophical riddles? Should we accept von Weizsäcker's double-aspect theory for psycho-somatic medicine?

Verwey argues convincingly *against* Reiner's causal-theory interpretation of von Weizsäcker's theory, and *for* the view that von Weizsäcker's position is a kind of double-aspect theory à la Fechner – both trying to reconcile the approach of natural science and a religious (hermeneutic) approach. Interpreted in that way, von Weizsäcker's psychosomatics is certainly not a contribution to solving the mind-body problem, because his basic idea of the 'subject-containing object' is not supported by theoretical models or arguments that display the structure of this unique ontological entity, but is rather a philosophical metaphor that indicates that attempts to reduce man to either a machine or disembodied spiritual entity are doomed to fail.

That mechanism and romantic spiritualism are both wrong or one-sided conceptions of human existence does, of course, not imply that there is a *riddle* about our bodily and spiritual dimensions, inadequately accounted for by the two positions in question, or beyond our cognitive grasp. It might as well be the case that the two conflicting positions and the dichotomy between the objective and the subjective implied by both is rooted in dubious philosophical assumptions.

In the complex philosophical landscape of the West, there are at least two other approaches to the philosophical perplexities concerning human consciousness: Hegel's and the late Wittgenstein's. From both positions the *subjective* as something qualitatively distinct from the *objective* (and material) is radically criticized, not by claiming that the mental is objective (or material) but by completely rejecting the dichotomy between the subjective and the objective (as accounted for in the Enlightenment).

Both positions reject the idea shared by various modern philosophies of mind – that we are immediately aware of our consciousness or of phenomena of consciousness. Both positions imply an abandonment of certain Cartesian conceptions (according to which the objects of our immediate awareness are phenomena of the soul) since we have an immediate knowledge of the unity of our mind and body.

In Hegelian and Wittgensteinian perspectives consciousness is always *mediated*, something which is brought off in a medium, through concepts imbedded in institutions, practices, and forms of life. The constituents of our mental life (our feelings, desires and ideas) are not (as stressed by Charles Taylor [6]), merely given as objects which surround us in the world as given. They are not data. In Taylor's terms: "we have to see self-perception as something we do, as something we can bring off, rather than as a feature of our basic predicament. This means that we see it as the fruit of an activity of formulating how things are with us .... In this way, grasping what we desire or feel is something we can all together fail to do, or do in a distorting or partial censored fashion" ([6] p. 85). Not only is our grasping mediated rather than direct and infallible, but also the features of ourselves that are grasped through self-perception are "themselves bound up with activity". This implies that "desires, feelings may not be understood as just mental givens, but as the inner reflection of the life processes we are" ([6], p. 86).

What are the implications of accounts of human consciousness (Hegelian or Wittgensteinian) for psychosomatics when we take *activity* as a basis for understanding our whole 'inner life'? First of all, it implies that there is no such thing as a philosophical riddle regarding the unity of mind and body. The idea of such a riddle presupposes that the mental and the bodily are two distinct kinds of phenomena, the unity of which can only be experienced but not explained. In the light of "activity theory" the unity of the mental and the physical is not something immediately experienced, but a conceptual matter. As the unity is not something *experienced* it cannot be a *mysterious*, unexplicable experience. On the contrary, the conceptual unity of mind and body can be accounted for by explicating how the unity is constituted phylogenetically and historically by showing what were the practices and forms of life which gave rise to particular bodily and mental structures, and how they constituted ontogenetically or biographically, i.e., in the life of particular persons.

Though there is no general philosophical riddle of the unity of the mental and the physical, there are an indefinite number of personal riddles, i.e., problems of how mental and physical states are integrated in the life-history of particular persons. So, our difficulties are not due to general philosophical impediments or mysteries, but to the consequences of the fact that forms of life and values (in the sense of concrete ideals or prototypes of action) are various and multiple.

Evidence is adequate to demonstrate that bodily and mental states are integrated in human diseases in complex ways. Recognizing this fact, but leaving it unexplained as a philosophical riddle, does not contribute to the growth of medical knowledge. However, by developing methods for understanding individual life-histories and forms of life (and by implementing such methods in education and clinical practice) we would significantly contribute to the growth of medical knowledge. An anthropological approach [4] that respects the perspective of the patient, and the patient's own (perhaps tacit) knowledge of his or her problems would provide medicine with conceptual and ethical resources that would enhance scientific and therapeutic progress. Implementing such an approach in medicine, however, would have far-reaching implications than merely adding new courses to medical curricula. It requires a new setting for clinical practice and health care, a context which institutionalizes respect for the patient, the patient's rights, perspective, and knowledge, and enables patients to participate in their treatment decisions and evaluation. As I stressed earlier, structural changes within our health-care system may be the most important step not only toward the growth of but perhaps also toward a revolution in medical knowledge.

*Aarhus University,*
*Aarhus, Denmark*

## BIBLIOGRAPHY

1. Jensen, U.J.: 1987, *Practice and Progress. A Theory for the Modern Health Care System*. Blackwell Scientific Publ., Oxford.
2. Jensen, U.J.: 1987, *Selvbedrag og Selverkendelse* [*Selfdeception and Self-knowledge*], Reitzel, Copenhagen.
3. Jensen, U.J.: 1989, 'From Good Medical Practice to Best Medical Practice', *International Journal of Health Planning and Management* 4(4), pp. 167–180.
4. Kleinman, A.: 1988, *The Illness Narratives. Suffering, Healing and the Human Condition*, Basic Books, New York.
5. Spicker, S.F.: 1990, 'Invulnerability and Medicine's "Promise" of Immortality: Changing Images of the Human Body during the Growth of Medical Knowledge', in this volume, pp. 163–175.
6. Taylor, C.: 1985, 'Hegel's Philosophy of Mind' in *Human Agency and Language, Philosophical Papers*, vol. 1. Cambridge University Press, Cambridge.
7. Thomas, L.: 1984, *The Youngest Science*. Oxford University Press, Oxford.
8. Verwey, G.: 1990, 'Medicine, Anthropology, and the Human Body', in this volume, pp. 133–162.

# NOTES ON CONTRIBUTORS

H. Tristram Engelhardt, Jr. , Ph.D., M.D., is Professor, Departments of Medicine and Community Medicine; Member, Center for Ethics, Medicine, and Public Issues, Baylor College of Medicine, Houston, Texas, U.S.A. He is also Professor, Department of Philosophy, Rice University, Houston, Texas, U.S.A.

Henk A. M. J. ten Have, M.D., Ph.D., is Professor, Department of Health Care Ethics and Philosophy, University of Limburg, Maastricht, The Netherlands.

Uffe Juul Jensen, Ph.D., is Professor, Department of Philosophy, Aarhus University, Aarhus, Denmark.

Gerrit K. Kimsma, M.D., is General Practitioner and Philosopher, Westzaan, The Netherlands.

B. Ingemar B. Lindahl, Dr. Med. Sc., is Associate Professor, Karolinska Institute, Department of Social Medicine, Huddinge University Hospital, Huddinge, Sweden.

Lennart Nordenfelt, Ph.D., is Professor, Department of Health and Society, Linköping University, Linköping, Sweden.

Stuart F. Spicker, Ph.D., is Professor, Department of Community Medicine and Health Care, School of Medicine, University of Connecticut Health Center, Farmington, Connecticut, U.S.A.

Paul Thung, emeritus Professor, Department of Metamedica, Medical Faculty of the University of Leiden, The Netherlands.

Gerlof Verwey, Ph.D., is Assistant Professor, Department of Philosophical Anthropology and Philosophy of Medicine, Catholic University of Nijmegen, The Netherlands.

Henrik R. Wulff, M.D., is Chief Physician, Department of Medicine, Herlev University Hospital, Herlev, Denmark.

# INDEX

*The Philosophy and Medicine Book Series*

*Editors*

H. Tristram Engelhardt, Jr. and Stuart F. Spicker

1. **Evaluation and Explanation in the Biomedical Sciences**
   1975, vi + 240 pp. ISBN 90–277–0553–4
2. **Philosophical Dimensions of the Neuro-Medical Sciences**
   1976, vi + 274 pp. ISBN 90–277–0672–7
3. **Philosophical Medical Ethics: Its Nature and Significance**
   1977, vi + 252 pp. ISBN 90–277–0772–3
4. **Mental Health: Philosophical Perspectives**
   1978, xxii + 302 pp. ISBN 90–277–0828–2
5. **Mental Illness: Law and Public Policy**
   1980, xvii + 254 pp. ISBN 90–277–1057–0
6. **Clinical Judgment: A Critical Appraisal**
   1979, xxvi + 278 pp. ISBN 90–277–0952–1
7. **Organism, Medicine, and Metaphysics**
   *Essays in Honor of Hans Jonas on his 75th Birthday, May 10, 1978*
   1978, xxvii + 330 pp. ISBN 90–277–0823–1
8. **Justice and Health Care**
   1981, xiv + 238 pp. ISBN 90–277–1207–7(HB) 90–277–1251–4(PB)
9. **The Law-Medicine Relation: A Philosophical Exploration**
   1981, xxx +•292 pp. ISBN 90–277–1217–4
10. **New Knowledge in the Biomedical Sciences**
    1982, xviii + 244 pp. ISBN 90–277–1319–7
11. **Beneficence and Health Care**
    1982, xvi + 264 pp. ISBN 90–277–1377–4
12. **Responsibility in Health Care**
    1982, xxiii + 285 pp. ISBN 90–277–1417–7
13. **Abortion and the Status of the Fetus**
    1983, xxxii + 349 pp. ISBN 90–277–1493–2
14. **The Clinical Encounter**
    1983, xvi + 309 pp. ISBN 90–277–1593–9
15. **Ethics and Mental Retardation**
    1984, xvi + 254 pp. ISBN 90–277–1630–7
16. **Health, Disease, and Causal Explanations in Medicine**
    1984, xxx + 250 pp. ISBN 90–277–1660–9
17. **Virtue and Medicine**
    *Explorations in the Character of Medicine*
    1985, xx + 363 pp. ISBN 90–277–1808–3
18. **Medical Ethics in Antiquity**
    *Philosophical Perspectives on Abortion and Euthanasia*
    1985, xxvi + 242 pp. ISBN 90–277–1825–3(HB) 90–277–1915–2(PB)
19. **Ethics and Critical Care Medicine**
    1985, xxii + 236 pp. ISBN 90–277–1820–2

20. **Theology and Bioethics**
*Exploring the Foundations and Frontiers*
1985, xxiv + 314 pp. ISBN 90–277–1857–1
21. **The Price of Health**
1986, xxx + 280 pp. ISBN 90–277–2285–4
22. **Sexuality and Medicine**
*Volume I: Conceptual Roots*
1987, xxxii + 271 pp. ISBN 90–277–2290–0(HB) 90–277–2386–9(PB)
23. **Sexuality and Medicine**
*Volume II: Ethical Viewpoints in Transition*
1987, xxxii + 279 pp. ISBN 1–55608–013–1(HB) 1–55608–016–6(PB)
24. **Euthanasia and the Newborn**
*Conflicts Regarding Saving Lives*
1987, xxx + 313 pp. ISBN 90–277–3299–4(HB) 1–55608–039–5(PB)
25. **Ethical Dimensions of Geriatric Care Value Conflicts for the 21st Century**
1987, xxxiv + 298 pp. ISBN 1–55608–027–1
26. **On the Nature of Health**
*An Action-Theoretic Approach*
1987, xviii + 204 pp. ISBN 1–55608–032–8
27. **The Contraceptive Ethos**
*Reproductive Rights and Responsibilities*
1987, xxiv + 254 pp. ISBN 1–55608–035–2
28. **The Use of Human Beings in Research**
*With Special Reference to Clinical Trials*
1988, xxii + 292 pp. ISBN 1–55608–043–3
29. **The Physician as Captain of the Ship**
*A Critical Appraisal*
1988, xvi + 254 pp. ISBN 1–55608–044–1
30. **Health Care Systems**
*Moral Conflicts in European and American Public Policy*
1988, xx + 368 pp. ISBN 1–55608–045X
31. **Death: Beyond Whole-Brain Criteria**
1988, x + 276 pp. ISBN 1–55608–053–0
32. **Moral Theory and Moral Judgments in Medical Ethics**
1988, vi + 232 pp. ISBN 1–55608–060–3
33. **Children and Health Care: Moral and Social Issues**
1989, xxiv + 349 pp. ISBN 1–55608–078–6
34. **Catholic Perspectives on Medical Morals**
*Foundational Issues*
1989, vi + 308 pp. ISBN 1–55608–083–2
35. **Suicicide and Euthanasia**
*Historical and Contemporary Themes*
1989, vi + 286 pp. ISBN 0–7923–0106–4